Research on Emulative Hybrid Assembled Precast Concrete Shear Walls

Zhangfeng Zhu · Zhengxing Guo · Lei Tang

Research on Emulative Hybrid Assembled Precast Concrete Shear Walls

东南大学出版社
SOUTHEAST UNIVERSITY PRESS

Springer

Zhangfeng Zhu
College of Civil Engineering
Nanjing Tech University
Nanjing, Jiangsu, China

Zhengxing Guo
School of Civil Engineering
Southeast University
Nanjing, Jiangsu, China

Lei Tang
Department of Infrastructure
Southeast University
Nanjing, Jiangsu, China

ISBN 978-981-96-9577-5 ISBN 978-981-96-9578-2 (eBook)
https://doi.org/10.1007/978-981-96-9578-2

Jointly published with Southeast University Press
The print edition is not for sale in China Mainland. Customers from China Mainland please order the print book from: Southeast University Press.
ISBN of the Co-Publisher's edition: 978-7-5641-9137-5

Translation from the Chinese Simplified language edition: "Zhuangpeishi Hunningtu Hunhe Lianjie Jianliqiang Yanjiu/装配式混凝土混合连接剪力墙研究" by Zhangfeng Zhu et al., © Southeast University Press 2020. Published by Southeast University Press. All Rights Reserved.

This Springer imprint is published by the registered company Springer Nature Singapore Pte Ltd.
The registered company address is: 152 Beach Road, #21-01/04 Gateway East, Singapore 189721, Singapore

Preface

The assembled monolithic precast concrete shear wall structure, based on the principle of "equivalent cast-in-place", closely parallels traditional cast-in-place structures in terms of bearing capacity, stiffness, ductility and energy dissipation. The analytical theory and design methods can also directly apply existing results from cast-in-place concrete structures.

Constrained by the current technological level of the industry and the initial stages of development in prefabricated concrete building structures in China, the assembled monolithic precast concrete shear wall structure, designed for extensive demand in cast-in-place applications, has undergone notable advancements. This pioneering method has been primarily driven by the necessity to address the rising demand in the residential building market and to accommodate the swift depletion of labor resources. Central to this evolution is the adoption of grout sleeve connection technology, grout anchored overlap connection technology, and post-cast concrete technique.

The research team, drawing upon the implementation of scientific research projects such as "Key Technologies Research of Prefabricated Concrete Shear Wall Structures in Assembled Buildings" (2011BAJ10B03) as part of the 12th Five-Year National Science and Technology Support Program, and "Efficient Construction and Acceptance Technology Research and Demonstration of Key Joint Connections in Prefabricated Buildings" (2016YFC0701703) under the 13th Five-Year National Key R&D Program, has been actively engaged in the research and development of assembled monolithic precast concrete shear wall structures. We have acquired a profound comprehension and holistic understanding of the technical principles, processes, and benefits associated with this technology. In response to challenges such as early concrete cracking, concentrated concrete failure at joint locations, and the defects of reinforcement connection technology within assembled monolithic precast concrete shear wall structures, we have introduced the prefabricated hybrid assembled connection shear wall technology. This approach integrates post-tensioning unbonded prestressing technology, local unbonded grouted lapped rebar technology, and overlapped enclosed stirrup technology, with the aim of enhancing the seismic performance of precast concrete shear walls. With the support of the National Natural

Science Foundation project "Seismic Performance Research of Novel Hybrid Precast Concrete Shear Walls" (51308289), our research team has conducted a series of experimental studies and theoretical analyses, achieving significant research findings.

This book is founded upon the research efforts and noteworthy accomplishments of the research team in recent years, concentrating on hybrid assembled precast concrete shear walls. It offers a systematic introduction and in-depth interpretation of this innovative technology, encompassing technical essentials, experimental studies, numerical analysis, design methods, and structural improvements. The content of the book is systematic, structurally complete, and articulates key points with clarity. It empowers readers to develop a comprehensive and accurate understanding of hybrid assembled precast concrete shear walls, serving as a cornerstone and point of reference for forthcoming research undertakings in the field.

This book is divided into 7 chapters, with the main contents as follows: Chapter 1 provides an introduction, systematically presenting the background, classification, and representative technical features of precast concrete shear walls. It elaborates on the status of assembled monolithic precast concrete shear wall technology in China and briefly discusses its development direction. Chapter 2 delves into the issues present in Chinese assembled monolithic precast concrete shear walls, introducing the concept, key parameters, and detailed construction of emulative hybrid assembled precast concrete shear walls. Chapter 3 focuses on seismic performance testing of emulative hybrid assembled precast concrete shear walls, considering the parameters such as the area of prestressed bars, prestressing force, local unbonded length of grouted bars, and axial compression ratio. It highlights experimental phenomena, analysis of test data, and the influence of various parameters on the seismic performance of specimens. Chapter 4 concentrates on numerical analysis on the seismic performance of emulative hybrid assembled precast concrete shear walls using ABAQUS software, detailing the specific simulation process and presenting finite element parametric analysis results based on the tests. Chapter 5 discusses the design methods of emulative hybrid assembled precast concrete shear walls, focusing on the calculation methods for the bearing capacity and stiffness of emulative hybrid assembled precast concrete shear walls based on experimental results and ABAQUS finite element parametric analysis results. Chapter 6 details the restoration force model of emulative hybrid assembled precast concrete shear walls, presenting the restoration force model suitable for elastic-plastic analysis under monotonic loading and cyclic loading based on experimental results and ABAQUS finite element parametric analysis. Chapter 7 addresses structural improvements of emulative hybrid assembled precast concrete shear walls based on OpenSees, discussing analysis work for structural improvements of emulative hybrid assembled precast concrete shear walls on the OpenSees platform as a necessary supplement and expansion of the completed experiments and ABAQUS analysis. It substantiates the feasibility and rationality of relevant construction improvement plans and provides efficient construction recommendations.

This book was written by Associate Professor Zhangfeng Zhu from the College of Civil Engineering at Nanjing Tech University, Prof. Zhengxing Guo from the School of Civil Engineering at Southeast University, and Associate Researcher Lei Tang from the Department of Infrastructure at Southeast University. The research results covered in this book were jointly completed by the authors' research team and graduate students. The members of the research team include Associate Professor Jiabin Liu and Dr. Dongzhi Guan from the School of Civil Engineering at Southeast University, and the graduate students include Yin Zhu, Yakun Li from Southeast University, Xulei Zang, and Fengge Shi from Nanjing Tech University. The research work received significant support from the partnering entity Jiangsu Zhongnan Construction Industry Group Co., Ltd., and sincere thanks are extended for their assistance throughout the research process.

The emulative hybrid assembled precast concrete shear wall is a new solution proposed by the research team to address the shortcomings of the existing "equivalent cast-in-place" precast concrete shear wall technology. Understanding and perception of this solution will continuously evolve and may even change as subsequent research progresses. Additionally, due to the authors' limited theoretical level and practical experience, there may inevitably be shortcomings or errors in the book, and readers are kindly requested to provide criticism and corrections.

The research involved in this book was funded by National Key Research and Development Program of China (No: 2016YFC0701703) and National Natural Science Foundation of China (No: 51308289). The authors wish to acknowledge the funding agency and members participating in the project for their support.

Nanjing, China
April 2020

Zhangfeng Zhu
Zhengxing Guo
Lei Tang

Competing Interests The authors have no competing interests to declare that are relevant to the content of this manuscript.

Contents

List of Figures

List of Tables

Chapter 1
Introduction

1.1 Overview of Precast Concrete Shear Wall Technology

Precast concrete (PC) technology is a method of constructing structures by prefabricating components in a factory and then assembling them on-site. It represents a crucial breakthrough for transforming and upgrading the construction industry, alongside cast-in-place concrete technology. Compared to cast-in-place concrete, precast concrete technology boasts several significant advantages, including reliable material performance, high product quality, noticeably shorter construction times, reduced environmental pollution, lower consumption of natural resources, efficient labor resource allocation, and cost-saving measures [1, 2]. However, it also has notable drawbacks, such as doubts about structural integrity and higher requirements for the precision, quality, and skill levels of prefabrication and assembly processes.

In recent years, China has vigorously promoted the modernization of the construction industry. The 2016 "Opinions on Further Strengthening Urban Planning, Construction, and Management" issued by the Central Committee of the Communist Party of China and the State Council proposed to "increase policy support," aiming for prefabricated buildings to account for 30% of new construction within approximately ten years [3]. The "Outline for the Modernization of the Construction Industry" issued by the Ministry of Housing and Urban–Rural Development in 2017 stated that the ratio of prefabricated buildings should reach 20% by 2020 and over5 0% by 2025 [4]. The continuous release of national and industry incentive policies indicates a favorable opportunity and superior policy environment for the development of prefabricated concrete structures in China.

At the same time, under the national commitment to a green, low-carbon, and sustainable development path, the traditional construction industry, characterized by high material waste, energy consumption, and environmental burdens, faces urgent tasks for transformation and upgrading. Prefabricated concrete structures, which can

Z. Zhu et al., *Research on Emulative Hybrid Assembled Precast Concrete Shear Walls*,
https://doi.org/10.1007/978-981-96-9578-2_1

effectively improve material utilization, reduce energy consumption, and significantly minimize onsite wet work, represent a vital solution for achieving the construction industry's goals of "energy-saving, water-saving, material-saving, land-saving," and environmental protection.

Furthermore, as China's demographic dividend gradually diminishes, the labor supply–demand relationship will change. There will continue to be a shortage of intermediate and high-skilled labor, represented by technicians and skilled workers, as well as a deficiency in the generally skilled workforce. On the other hand, labor costs will continue to rise, and the construction industry, which relied on inexpensive labor or was labor-intensive, will face increased labor costs and shortages. This situation is prompting many enterprises to urgently seek more efficient and intensive industrialized construction production methods, providing a favorable opportunity for the development of prefabricated concrete structures suited to industrialization in construction.

As urbanization accelerates in China and housing supply becomes increasingly tight, the integration of prefabricated concrete structures with the shear wall structures commonly used in residential buildings has resulted in precast concrete shear wall structures, becoming one of the important solutions to the current significant gap in the housing market.

It is evident that under the combined influence of national policy incentives, industry development trends, and the practical needs of national construction, the utilization of PC construction structures employing PC technology aligns with China's sustainable development strategy. This approach will effectively promote the adjustment and upgrading of industrial structures, significantly enhance industry technological standards, and rapidly elevate the quality of industry professionals, garnering widespread attention and focused development within the industry. The integration of PC structures with the commonly used shear wall structures in residential buildings to form PC shear wall structures can effectively address the growing housing supply shortage, the increasing scarcity of labor resources, and the rapid rise in costs in China. As a result, it holds significant research value and promising application prospects.

Based on different seismic resistance mechanisms, PC shear walls can be categorized as fully assembled precast concrete shear wall and assembled monolithic precast concrete shear wall. Fully assembled precast concrete shear walls leverage the advantages of PC technology, often connecting prefabricated shear walls into a unified whole using "dry" connection methods such as post-stressed connections and bolted connections. The stress mechanism, analysis theory, design methods, and construction techniques of fully assembled precast concrete shear walls differ fundamentally from those of traditional cast-in-place concrete shear walls, resulting in the development of a novel structural system. Assembled monolithic precast concrete shear walls adopt the "equivalent cast-in-place" concept, emphasizing the need for reliable connections to ensure that the load-bearing capacity, stiffness, ductility, and energy dissipation capabilities of PC structures achieve a level of equivalence to cast-in-place concrete structures. This concept is derived from ACI550.1R-09 [5]'s "equivalent cast-in-place detailing" methodology, which addresses the connection issues of load-bearing reinforcement, while the concrete joint areas of PC components only consider the transfer of internal forces through compression, maintaining consistency with the theoretical assumptions and design methods of traditional cast-in-place concrete structures. Thereby, the approach establishs close integration with traditional cast-in-place concrete structures.

A series of technical research and development work has been carried out both domestically and internationally for fully assembled precast concrete shear wall and assembled monolithic precast concrete shear wall. Numerous experimental results have been achieved, leading to the formation of multiple representative technical systems. The details are as follows:

1.1.1 Fully Assembled Precast Concrete Shear Wall

In the development of PC shear wall technology, to maximize the technical benefits of PC, the fully assembled precast concrete shear wall primarily emphasizes the exploration and advancement of post-stressed connection technology. Concurrently, a variety of innovative technologies based on bolted connection technology and other available technologies sporadically emerge.

(1) Fully assembled precast concrete shear wall utilizing post-stressed connection technology

The research on post-stressed PC shear walls originated from the collaborative project PRESSS (Precast Seismic Structural Systems), initiated by the United States and Japan in 1990s [6]. In pursuit of a multi-story prefabricated structural system suitable for seismic zones, a seismic performance test was conducted on a five-story prefabricated building at a 60% scale. Post-stressed connections were applied to join PC shear walls to form a fully assembled precast concrete shear walls featuring their own unique stress characteristics. Test outcomes aligned with the design philosophy. The post-stressed connection concentrates the non-linear deformation in the connection region, leaving the remaining parts relatively undamaged or merely elastically deformed, necessitating only localized repairs post-earthquake. The elastic nature of the prestressing bars facilitates sufficient elastic recovery force during unloading, pulling the specimens back to their original position, minimizing residual deformation during unloading, and establishing the self-centering property [7].

Subsequently, post-stressed connection has evolved into a crucial technical approach embraced by fully assembled precast concrete shear walls. Referred to as the "rocking wall" by some scholars, the post-stressed PC shear wall has undergone ongoing enhancements in subsequent research endeavors.

Soudki and others [8] conducted a series of comparative experimental studies on the vertical connection nodes of post-stressed PC shear walls. These nodes were categorized based on the post-stressed steel used and their bonding status, including bonded post-stressed steel strand connections, bonded post-stressed steel bar connections, and unbonded post-stressed steel bar connections. The test results revealed that each joint possessed sufficient bearing capacity and ductility. Furthermore, owing to the self-recovery characteristics of the post-stressed bars, the energy consumption capacity of the specimens was moderately reduced, with the specimen utilizing post-stressed steel strands displaying the lowest energy consumption. Additionally, local unbonding of the post-stressed bars enhanced joint deformability and ensured continuous elasticity of the post-stressed bars.

Perez and colleagues [9–11] carried out a theoretical analysis of the whole stress process of the unbonded post-tensioned precast concrete shear wall assembly with vertical ductile connection and the separate unbonded post-tensioned precast concrete shear wall under horizontal monotonous load. They identified five limit states: Decompression at the wall base (DEC), Yielding of Vertical Joint Connectors (LLJ), Effective Linear Limit (ELL), Yielding of the PT steel (LLP), and Base Shear Capacity, Crushing of confined concrete at the wall base (CCC). They further simplified these concepts into a tri-linear theoretical specimen that reflects the entire loading process (ELL, LLP as linear inflection points), showing good consistency with experimental results. See Fig. 1.1.

Based on the test and theoretical research results of the post-stressed connection shear wall technical system, the American Concrete Society issued Acceptance Criteria for Special Unbonded Post-tensioned Precast Structural Walls Based on Validation Testing (ITG 5.1–07) [12] and Requirements for Design of a Special Unbonded Post-Tensioned Precast Shear Wall Satisfying ACI ITG-5.1 (ITG 5.2–09) [13], in order to guide and promote the engineering application and follow-up research of the post-stressed connection shear wall technology.

In order to overcome the problem of poor energy dissipation capacity of post-stressed connection shear wall, many scholars have carried out special research. Some scholars have proposed introducing additional energy dissipation components into post-stressed PC shear walls, including friction dampers, fluid dampers,

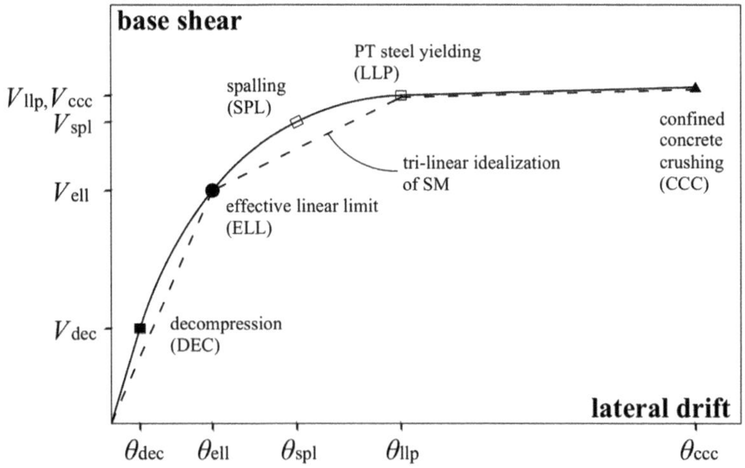

Fig. 1.1 Trilinear theoretical test piece of post-stressed connection shear wall [11]

and yielding energy-dissipation dampers, to compensate for their lack of energy-dissipation capacity while enhancing the integrity between wall segments and reducing horizontal displacement [14–18].

Hamid et al. [19] effectively combined the post-stressed connection technology with the pre-fabricated hollow wall panel technology and carried out experimental research on its combination components. The assembly consists of 6 panels, 2 panels at both ends are load-bearing wall panels and 4 panels at the middle are non-load-bearing wall panels. The load-bearing wall panels are connected to the base by prestressing compression, while the non-load-bearing wall panels are only placed on the base by rubber blocks, which form a two-point connection and fill with sealant. Among them, the post-stressed bars are constructed with threaded bars and variable diameter sections are set in the upper and middle part of the bars as "fuses". Variations of parameters such as diameter of prestressing bar, length of fuse, spacing between rubber blocks and sealant or not were considered in the test. The test results show that the assembled concrete hollow wall-slab assembly has good lateral resistance and has no obvious structural damage under the load of < 4% horizontal displacement.

Kurama [20] proposed a hybrid precast wall system (see Fig. 1.2a for the structural scheme, i.e. on the basis of prestressing connection, bonded mild steel bars are introduced, and two kinds of materials, prestressing reinforcement and bonded mild steel bars, are used for "mixing" to improve the energy dissipation capacity of shear wall by yield of bonded mild steel bars. In the mixed connection shear wall, the prestressing reinforcement usually uses the prestressing steel strand harness with unbonded length. The prestressing force formed by the tensioning prestressing reinforcement and the gravity of the member itself provide the self-recovery ability. The bonded mild steel bars crosses the joint at the end of the wall limb so that it can yield under tension and compression to dissipate energy and make up for the deficiency of energy dissipation capacity of pre-stressed connection shear wall. Smith et al. [21] further improved the mixed-connection shear wall. See Fig. 1.2b for the improvement plan. Move the bottom bonded mild steel bar to the middle of the wall limb. The other bonded mild steel bar is still set at the end of the wall limb. At the same time, the bonded mild steel bar has set a certain length of non-bonded section at the joint location. The bottom bonded mild steel bar still exerts its yield energy dissipation capacity, while the other upper bonded mild steel bar is not allowed to yield, so as to eliminate the crack opening and shear slip at the upper horizontal joint position. In addition, the form of end reinforcement of wall limbs is changed from circular screw reinforcement to rectangular Enclosed Reinforcement to increase the core area of concrete. Relevant studies have proved that: (1) The main mode of horizontal deformation of well-designed mixed- precast concrete shear wall is the opening of gaps at the joints. Relatively speaking, the horizontal displacement components of bending and shear deformation of the wall itself are small. The shear wall basically rotates around the joint as a rigid body and its damage degree is obviously lower than that of cast-in-place shear wall. (2) During unloading stage, the pre-stressed reinforcement relies on its elastic deformation to provide vertical recovery force to close the gaps open at the joints again, thus reducing the residual deformation of the members after earthquake. On the one hand, the use of unbonded PT tendon reduces the strain of the prestressing reinforcement itself and keeps it elastic at all times to provide sufficient elastic resilience; On the other hand, it cuts off the transmission of tensile stress from the pre-stressed reinforcement to the surrounding concrete, thus avoiding the cracking of concrete to a certain extent. At the same time, unbonded PT tendon is usually set near the middle of the wall limb, which can shorten the extension of the prestressing reinforcement in the loading process as far as possible, and also make the prestressing reinforcement far away from the confined concrete at the end of the wall limb, thus delaying the collapse of the confined concrete. (3) The yield of local unbonded ordinary steel bars is obviously lagged and low-cycle fatigue breaking is avoided. The energy consumption capacity of members is significantly increased by the tensile and compressive yield of bonded mild steel bars. Moreover, the existence of bonded mild steel bars effectively restrains the shear slip of members along horizontal joints and improves the hysteretic performance of members.

Sritharan et al. [22] proposed the Precast Wall with End Columns technology, see Fig. 1.3. The system additionally establishes end column (CFT/Steel/Concrete) at the ends of precast walls, which are connected to the precast walls by vertical energy dissipators in the form of oval steel belts (o-connectors). Relevant test results show that elliptical steel strips can play a good role in connection and energy consumption. Prefabricated wall panels are only slightly damaged, concrete protective layer is only peeled off near the base, and end columns are not significantly damaged, showing good overall seismic performance.

In recent years, some domestic scholars have made follow-up research on fully assembled precast concrete shear wall based on pre-stressed connection technology [23, 24], which has done pioneering work for further research and subsequent application of pre-stressed connection shear wall technology in China.

(2) Fully assembled precast concrete shear wall based on bolted connection technology

The most famous fully assembled precast concrete shear wall based on bolted connection technology is the bolted connection shear wall technology of Peck Group of Finland [25]. It has been commercialized successfully and has actively developed its application market in China in recent years.

Bora et al. [26] proposed a steel plate bolted connection shear wall connection schematic as shown in Fig. 1.4. The connection structure can restrain excessive vertical internal force by sliding bolts in the vertical groove and provide certain vertical resistance by initial friction. It can be used to connect thinner pre-stressed concrete wall slabs or pre-fabricated hollow wall slabs to the foundation to avoid anchorage failure of the connection structure and brittle failure of the member itself.

Semelawy et al. [27] proposed the technology of bolted connection shear wall with stiffening ribs. The conceptual schematic diagram is shown in Fig. 1.5. Each prefabricated wall panel is reinforced by four vertical stiffening ribs. Bolt holes are provided in the protruding space between the stiffening ribs and the prefabricated wall panel. Bolt holes are also provided along the vertical direction for both sides of the stiffening ribs. The prefabricated wall panel can be assembled and connected

Fig. 1.2 Mixed connection shear wall system

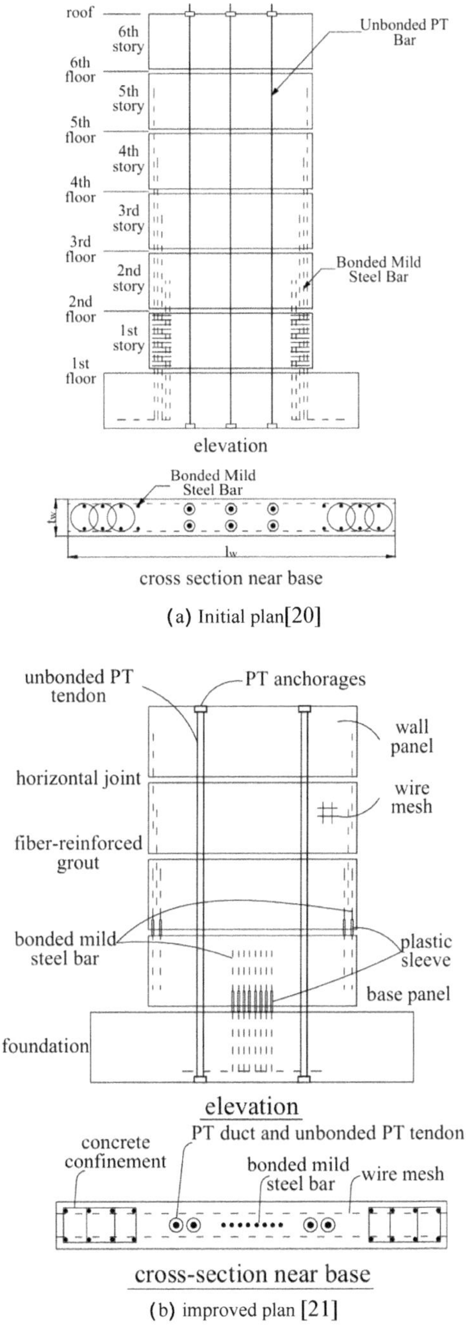

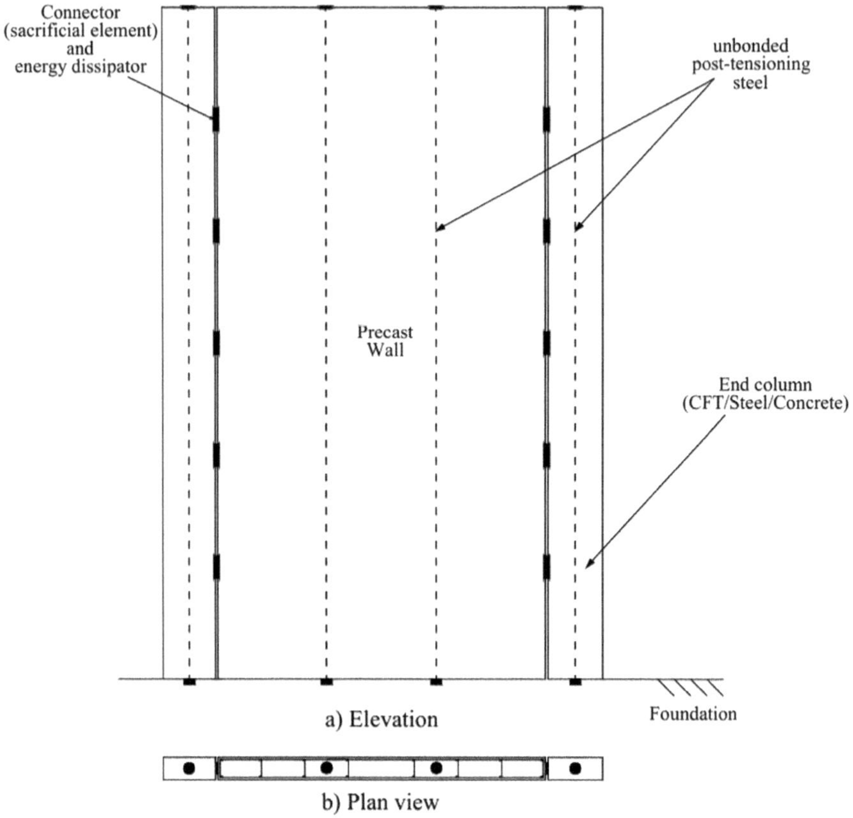

Fig. 1.3 PreWEC technology [22]

vertically and horizontally by bolts to resist both horizontal loads and shear deformation between the prefabricated wall panels caused by horizontal loads. The test shows that the plastic deformation concentrates on the horizontal joint, which shows the opening and closing of the joint. Plastic strain concentrates on bolts, but has little effect on concrete. It can be designed as a replaceable scheme, i.e. only bolts need to be replaced after earthquake. However, this scheme can only be used in multi-storey building structures with medium and low seismic fortification intensity.

For Peck bolt connection technology and bolt connection technology based on dark beam, tracking research has also been carried out in China, and a large number of test data results have been obtained [28, 29].

(3) Fully assembled precast concrete shear walls based on other available technologies

Dusicka et al. [30] proposed an innovative ICF (Insulated Concrete Form) grille shear wall system (see Fig. 1.6). It is spliced by standard ICF modules, which form reinforced concrete grilles (R/C Cores) in cavities. ICF works with grilles to provide

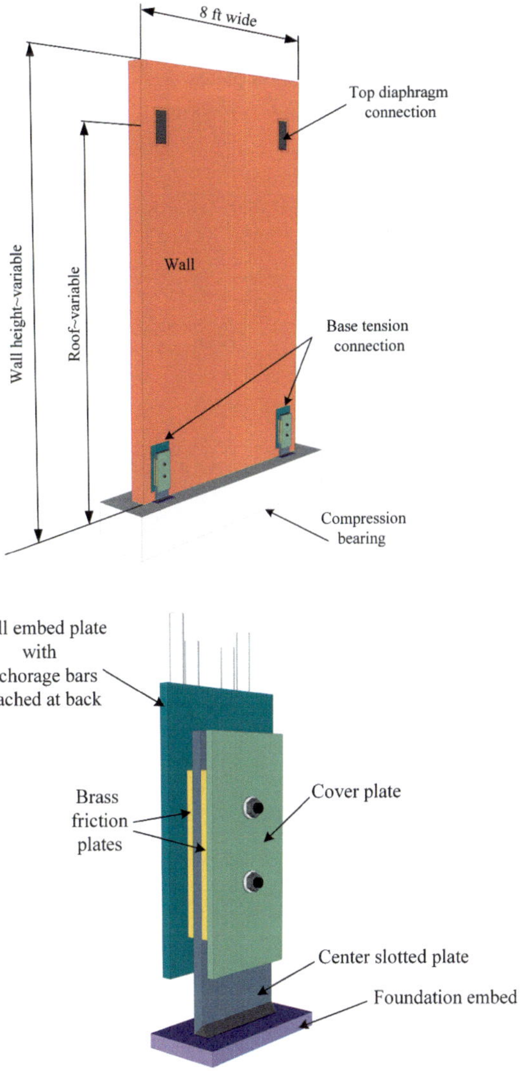

Fig. 1.4 Schematic diagram of steel plate bolted connection shear wall connection [26]

the lateral bearing capacity of shear wall components. The test results show that even if the strength of ICF material is only 1% of that of grille concrete, it is beneficial to improve the side resistance of the integral wall.

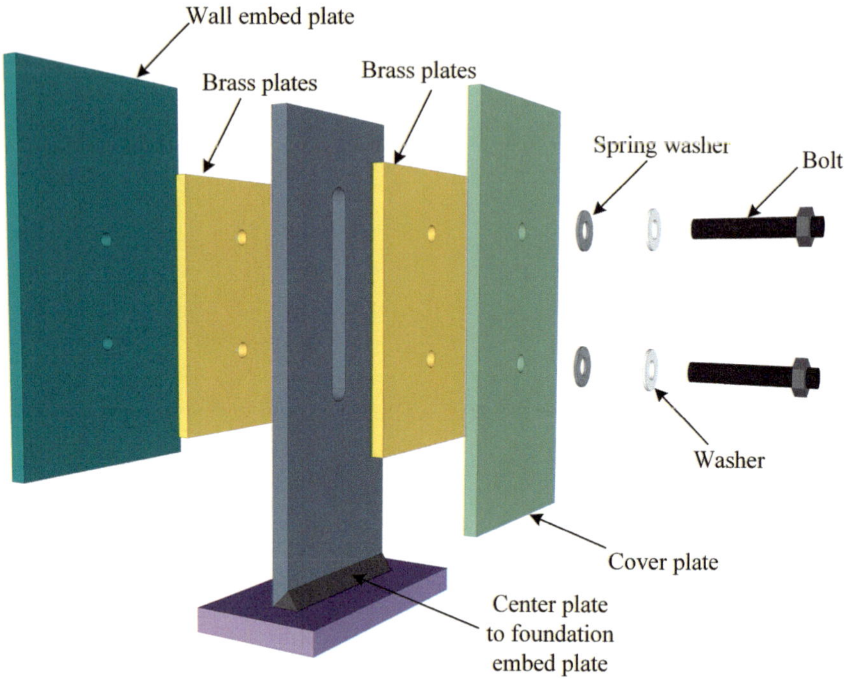

Fig. 1.4 (continued)

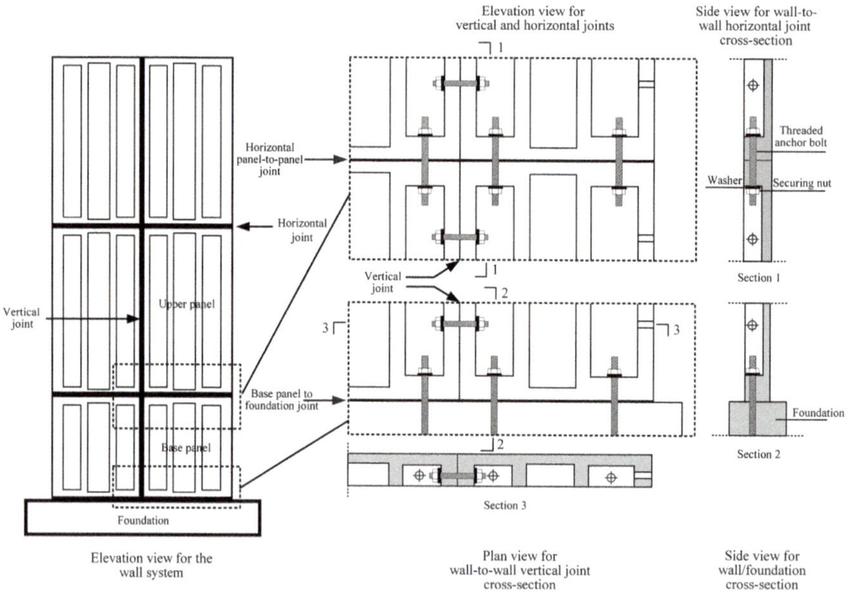

Fig. 1.5 Technical concept sketch of bolted connection shear wall with stiffening ribs [27]

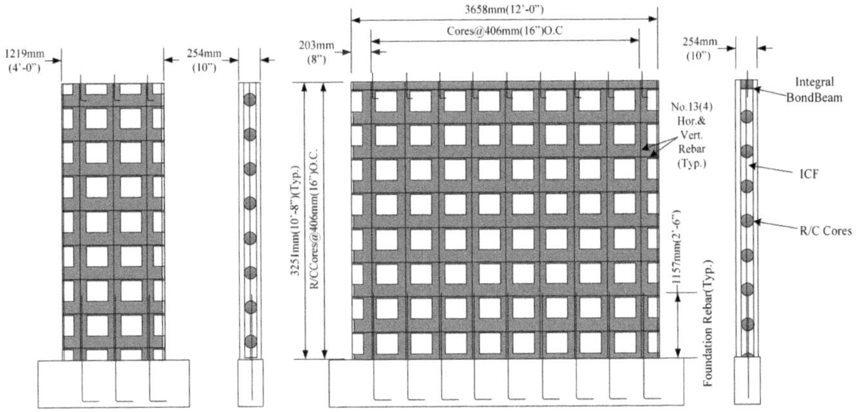

Fig. 1.6 ICF grille wall [30]

1.1.2 Assembled Monolithic Precast Concrete Shear Wall

In order to achieve the same integrity as cast-in-place concrete shear wall, assembled monolithic precast concrete shear wall is analyzed from the point of view of design concept, as long as the problem of continuous force transmission of reinforcing bars at joints formed between prefabricated wall panels is solved, that is, the problem of connection of reinforcing bars. To solve this problem, assembled monolithic precast concrete shear wall based on reinforcement connection technology and superimposed monolithic concrete shear wall based on core concrete post-pouring are formed. During the development of assembled monolithic precast concrete shear walls, other innovative and obvious technologies emerged sporadically.

(1) Assembled monolithic precast concrete shear wall based on steel bar connection technology

As mentioned earlier, the concept of "equivalent cast-in-place" was first proposed by the American Concrete Society in ACI 550.1R-09 [5]. In this report, an assembled monolithic precast concrete shear wall structure based on steel bar overlap, mechanical connection of steel bar and sleeve grout anchor connection is presented, as shown in Fig. 1.7. The steel bar overlap and mechanical connection both need to reserve high vertical passage in the wall panel. The lower prefabricated steel bar extends and enters the upper wall panel passage. The corresponding steel bar overlaps or mechanically connects with the upper wall panel in the passage, and then fills the high-strength grout mixture to form the integral connection. Sleeve grout anchor connection needs special sleeve for embedded steel bar connection in prefabricated wall panel (divided into full grouting sleeve with both ends connected with steel bar grouting anchor and half grouting sleeve with one end connected with steel bar grouting anchor and one end connected with steel bar threads). After steel bar is inserted into sleeve, the high-strength grouting material mixture is poured to form an integral connection.

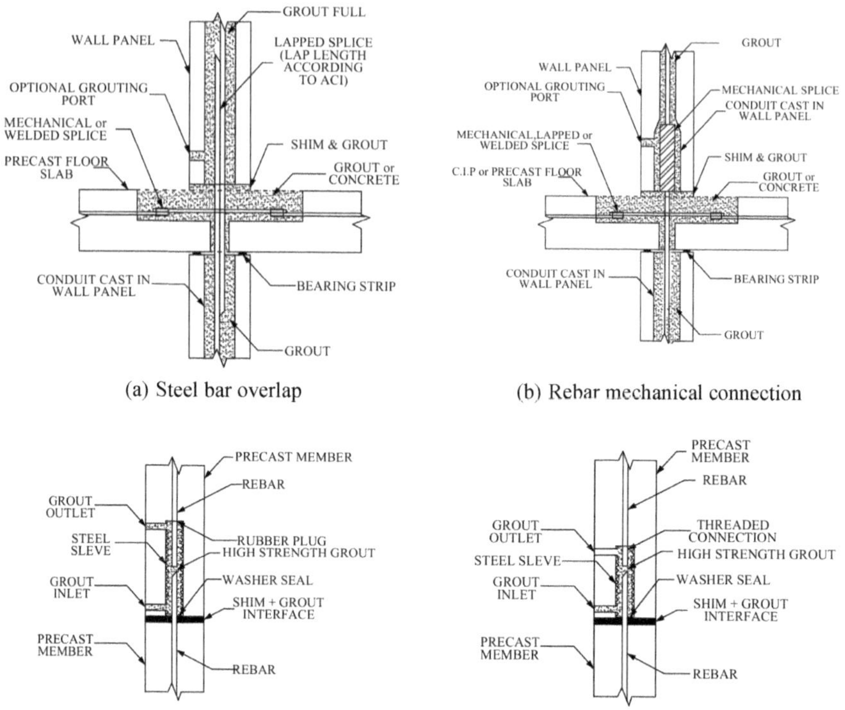

(a) Steel bar overlap (b) Rebar mechanical connection

(c) Fully sleeve grout anchor connection (d) Partially sleeve grout anchor connection

Fig. 1.7 ACI 550.1R-09 recommended construction [5]

As the technology of assembled monolithic precast concrete shear wall based on reinforcement connection technology is simple and basically the same as that of cast-in-place concrete structure, it is generally recognized and paid attention in China. In recent years, a lot of research achievements have been achieved and large-scale popularization and application has been carried out.

(2) Composite integral concrete shear wall based on core concrete post-pouring

According to the different ways of overlapping the integral concrete shear wall, it can be divided into two-sided overlapping shear wall and single-sided overlapping shear wall.

Double-wall Precast Concrete Building System in Germany is the representative of the double-wall composite shear wall. The composite wall slab consists of two reinforced concrete slabs at least 5 cm thick and firmly connected by truss reinforcement. When the composite wall slab is installed in place, self-compacting concrete is poured between the two prefabricated slabs. The typical picture of the joint of the wall slab is shown in Fig. 1.8a. The two prefabricated wall panels can be used both as load-bearing and as side formwork of cast-in-place concrete. Prefabricated laminated floor slabs are the same as pre-fabricated laminated wall slabs except for the

(a) Typical node (b) Prefabricated floors

Fig. 1.8 German double-wall technology [30]

different thickness and reinforcement. At the same time, polyurethane foam panels can be filled between the upper and lower prefabricated panels to form a sandwich type to reduce floor dead weight and improve sound insulation effect, as shown in Fig. 1.8b.

For the fabrication of prefabricated overlapping wall panels, Germany has an advanced equipment and fabrication and maintenance process. The fabrication process is as follows: firstly, one side wall panel (Side A) is fabricated, and then it should be fabricated together with lattice reinforcement; secondly, the other side wall panel (Side B) is poured. Before the initial setting of concrete and when the demoulding strength of Side A concrete is reached, the Side A is absorbed and lifted by suction cup and pressed vertically into the Side B concrete to the specified depth to ensure that the thickness of the composite wall plate and the protective layer of reinforcement meet the design requirements. Finally, it is sent to the three-dimensional steam curing room for steam curing. The detailed process can be seen in Fig. 1.9.

Single-sided overlapping shear wall is represented by Precast Concrete Form technology imported from Japan by Vanke Group. Its essence is prefabricated concrete formwork technology, i.e. the problem of supporting formwork on outer wall panel is solved by prefabrication of 50 mm thick outer wall panel, avoiding the support of scaffolding and formwork on outer wall panel. Pictures of typical PCF construction are shown in Fig. 1.10.

Due to the advanced production process and equipment, the technology of superimposed monolithic concrete shear wall has been actively introduced and specially studied in China in recent years, resulting in several production lines with high degree of automation and a lot of research results.

(3) Other assembled monolithic precast concrete shear walls

Holden et al. [31] proposed a simulated cast-in-place assembled shear wall unit (see Fig. 1.11). The prefabricated shear wall is inserted into the reserved pit of the base (the depth of the pit is equal to the length of the vertical reinforcement anchorage of the prefabricated shear wall). Transverse deformed reinforcement bars are set along the length of the wall at the corresponding positions of the wall and the base and the gaps

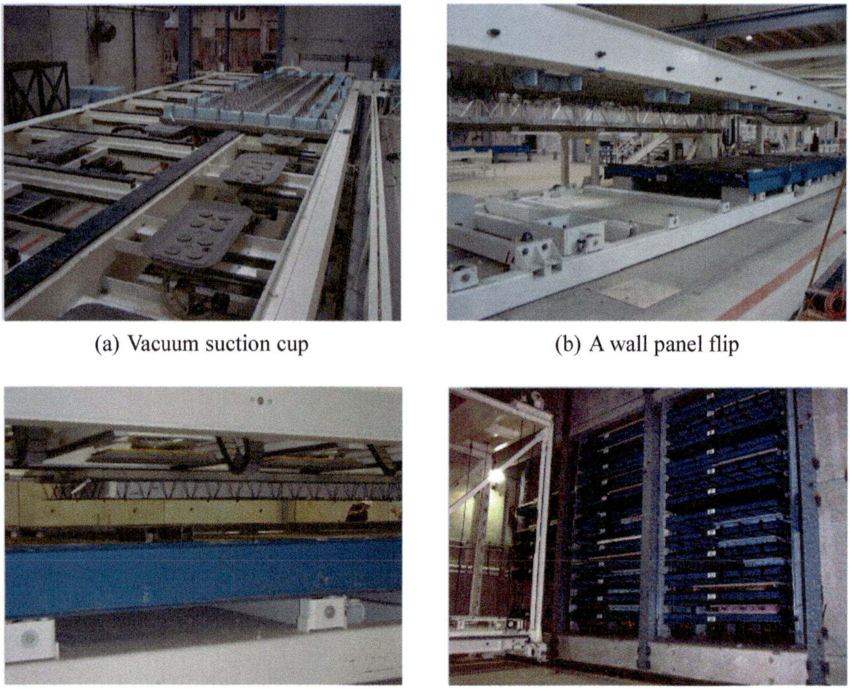

(a) Vacuum suction cup

(b) A wall panel flip

(c) A wall panel is buckled into B wall panel

(d) Steam Maintenance

Fig. 1.9 German double-wall prefabrication technology [30]

Fig. 1.10 PCF technical construction photos

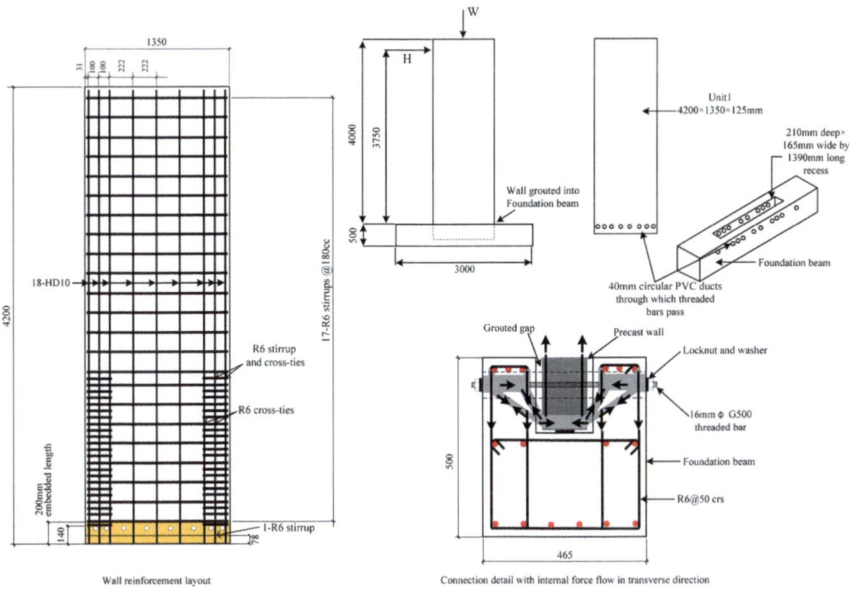

Fig. 1.11 A simulated cast-in-place assembled shear wall unit

are filled with high-strength grouting material. Relevant seismic performance test results show that the connection structure of the simulated cast-in-place assembled shear wall unit can guarantee the integrity of the prefabricated shear wall and the base, and the plastic hinge at the foot of the wall is fully developed, basically equal to the cast-in-place shear wall, but the residual deformation of the specimen is large and the damage is serious.

In addition, a Singapore project employs prefabricated wall panel connection technology as shown in Fig. 1.12. The vertical connection of shear wall is strengthened by welding reinforcement cages with screw bars at the end of shear wall and several vertical small reinforcements to reinforce the connection of grout pillars and the pier-head coarse reinforcement projecting from the upper and lower walls. The analysis shows that the connection mainly relies on the upper and lower coarse steel bars with pier head to transmit internal force, at the same time, the welding reinforcement cage restrains grouting material, increases its strength and grip force to the coarse steel bars, thus guaranteeing the bearing capacity requirements. However, the stress of this connection is similar to that of short columns, and the grouting material is more brittle than concrete. Therefore, its ductility performance needs to be discussed. At the same time, considering that Singapore is predominantly a wind load control design, the connection form should be sufficient to meet the stress requirements.

(a) Prefabrication of shear wall panels (b) Prefabricated Wall Panel Stacking

(c) Shear Wall Connection Rough and Spiral Bars (d) Shear Wall Installation

Fig. 1.12 Technology used in a Singapore project

1.2 Development Status of Precast Concrete Shear Wall in China

In recent years, the state has vigorously advocated the modernization of the construction industry, and strived to promote the transformation and upgrading of the construction industry through the development of prefabricated buildings, and to achieve green, low-carbon and sustainable development of the construction industry. Therefore, prefabricated concrete buildings, especially to meet a large number of large-scale residential buildings in China precast concrete shear wall structure technology has been rapid development. However, due to the late start of the research and development of prefabricated buildings in China, it is still in the primary stage on the whole, which is restricted by the technical level of the industry. At the same time, considering the realistic requirements of the development goals in the short term, the assembled monolithic precast concrete shear wall structure that can basically achieve' seamless docking' with the cast-in-place concrete structure has received widespread attention and key development.

At present, the development of prefabricated concrete shear wall technology in China mainly focuses on two directions: assembled monolithic precast concrete shear wall and superimposed monolithic shear wall based on steel bar connection technology. At the same time, innovative fully assembled precast concrete shear wall technology has emerged sporadically.

(1) Development status of assembled monolithic precast concrete shear wall based on steel bar connection technology in China

With the "*Technical specification for precast concrete structures*" *(JGJ 1-2014)* [32] and "*Technical standard for assembled buildings with concrete structure*" *(GB/T 51231-2016)* [33] have been promulgated and implemented, China has basically formed a vertical reinforcement connection technology as the core of the assembled monolithic precast concrete shear wall structure technology system, and roughly divided into reinforcement grout sleeve connection technology and reinforcement grout anchored overlap connection technology, namely the upper and lower precast shear wall vertical reinforcement connection through reinforcement grout sleeve connection technology or reinforcement grout anchored overlap connection technology to achieve (see Fig. 1.13). The adjacent precast shear walls on the same floor are connected by post-cast concrete, and the vertical members and horizontal members are connected into a whole through the cast-in-place concrete of the composite floor. Among them, the reinforcement grout sleeve connection technology is the introduction of foreign mature technology. Through the independent research and development of sleeve and grouting material products, it has formed its own products, such as JM grouting sleeve and CGMJM-VI type joint special grouting material developed by Beijing Jianmao Construction Equipment Co., Ltd., Shenzhen Modern Construction Technology Co., Ltd. developed " concrete" ductile iron semi-grouting sleeve, Taiwan Runtai Group developed full grouting sleeve made of ductile iron casting and Liuzhou Ouweim Machinery Co., Ltd. developed GT-type steel bar connection grouting sleeve system for assembled bridges. According to the difference of hole forming method, joint constraint structure and grouting method, the reinforcement grout anchored overlap connection technology is divided into reserved hole grout anchored overlap connection (constrained slurry anchor) and corrugated pipe grout anchored overlap connection (unconstrained slurry anchor). The former improves the joint performance by core-pulling hole forming and adding spiral reinforcement constraint at the joint and adopts pressure grouting process. The latter improves the joint performance by pre-embedded metal bellows hole forming and relying on horizontal reinforcement or edge component stirrup at the joint, and can choose pressure grouting or gravity grouting process.

Reinforcement grout sleeve connection technology can realize the axial force transmission of steel bars and meet the requirements of mechanical connection of steel bars. It has become the main steel bar connection technology in China. However, reinforcement grout anchored overlap connection, due to its eccentric force transmission, has certain limitations in its application scope, usage locations, and suitable working conditions, often requiring experimental verification of mechanical performance and applicability.

Based on the research results of the past ten years, with the continuous improvement of the prefabricated concrete construction technology in the industry, the

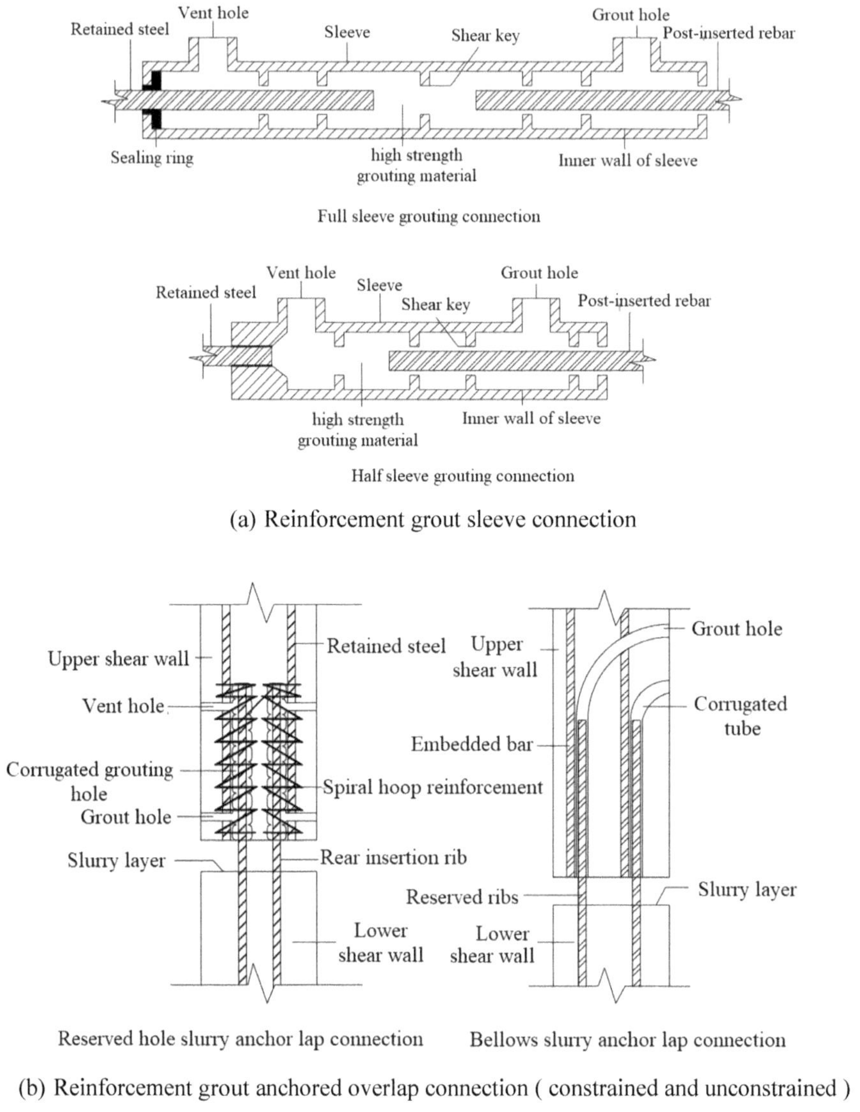

Fig. 1.13 Vertical reinforcement connection technology of assembled monolithic precast concrete shear wall in China

seismic performance of the assembled monolithic precast concrete shear wall structure in China has been widely recognized. Its important market application index is that the maximum applicable height of the house is basically the same as the cast-in-place concrete structure under its applicable seismic fortification intensity conditions. The relevant specifications in China are shown in Table 1.1. For the assembled monolithic precast concrete shear wall structure, when the total shear force at the bottom of the precast shear wall member is greater than 80% of the total shear force of the layer, the maximum applicable height should be taken as the value in the brackets in Table 1.1.

In addition, in view of the practical problems faced by grout sleeve connection technology, such as high requirements for prefabrication and installation, uneven grouting quality and difficulty in detection, some scholars have carried out relevant improvement attempts, including the use of non-uniform distribution of large-diameter steel grout sleeve connection to reduce steel joint [34, 35]; the grout sleeve connection is replaced by the joint connection beam [36], the ring bar buckle anchor connection [37, 38] or the embedded steel plate bolt connection [39] in the horizontal seam part. Or the prefabricated wallboard with cavity is used to provide conditions for the lapping connection of vertical reinforcement in the cavity to avoid the use of grout sleeve connection [40, 41].

(2) Development status of composite integral concrete shear wall in China

The technology of composite integral concrete shear wall in China is also developing rapidly. Based on the introduction of Japanese PCF technology and German "Double-Wall Precast Concrete Building System" technology, the localization research of this kind of technology is carried out. For example, Xing et al. [42] carried out low cyclic loading test and nonlinear analysis of composite slab shear wall, and compared the bearing capacity, hysteresis performance and failure mode of the structure. It is considered that there is no significant difference in the seismic performance of precast composite slab shear wall under different edge constraint construction measures. Based on the improvement of the component structure and connection mode of the original Double-Wall system, Quandong [43] carried out low cyclic loading tests on the I-shaped double-sided composite shear wall specimens with different edge structures and different shear span ratios, and evaluated the seismic performance of the specimens such as hysteresis curve, skeleton curve, displacement ductility, stiffness degradation and energy dissipation capacity. Zhang et al. [44] studied the nonlinear performance of single-sided composite shear wall. Relevant research shows that based on the advanced automatic flow production line of laminated wallboards abroad, structural improvements are made to meet the requirements of China's seismic design, such as the use of stirrups instead of truss steel bars in the edge member or the cast-in-place form of the edge member. Such imported technology can meet the seismic requirements of our country and has the application conditions.

Table 1.1 The maximum applicable height of assembled monolithic shear wall structure in China (m)

Structure type	Earthquake fortification intensity				
	6°	7°	8° (0.2 g)	8° (0.3 g)	9°
Cast-in-place structure	140	120	100	80	60
Prefabricated monolithic structure	130 (120)	110 (100)	90 (80)	70 (60)	–

Because of its technical characteristics, the composite integral shear wall is not valued in the early stage of development in China, and is mostly used in underground garage structure. In recent years, Meihao Group and the German AbbVie Group reached a cooperation, the introduction of a number of double-sided composite wall automation production line; on the basis of the original double-sided composite wall technology, SANY CONSTRUCTION has carried out systematic improvement and formed the assembled monolithic welded steel mesh laminated concrete structure technology (SPCS structure system) with independent intellectual property rights. Many places have invested in the research and development of composite integral shear walls, forming multiple local standards. It can be seen that after years of development and precipitation of prefabricated building technology in China, the advantages of composite integral shear wall have been re-recognized by the industry and will soon usher in a new development climax.

(3) Development status of fully assembled precast concrete shear wall in China

While the assembled monolithic precast concrete shear wall has been intensively developed, some domestic scholars have also carried out research work on the fully assembled precast concrete shear wall technology. In addition to the follow-up study of post-stressed connection shear wall and bolted connection shear wall [23, 24, 28, 29], innovative fully assembled precast concrete shear wall technology has also been proposed. For example, the post-welded steel plate shear connection key (SSK) is used for the vertical joint of the precast shear wall to reduce the relative deformation of the wall on both sides of the joint [45], or the precast shear wall is connected by the embedded border of wallboard, horizontal connection steel frame and high-strength bolt to avoid the wet operation of the joint. Improve the construction speed [46], related structure diagram 1.14 (Fig. 1.14).

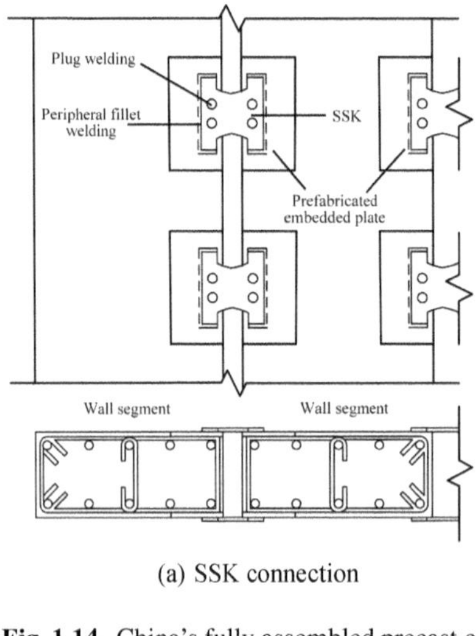

(a) SSK connection

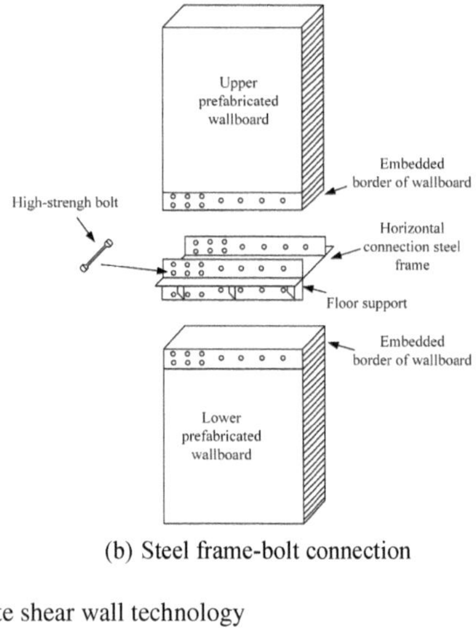

(b) Steel frame-bolt connection

Fig. 1.14 China's fully assembled precast concrete shear wall technology

At present, China's fully assembled precast concrete shear wall technology is still in its infancy, but with the continuous deepening of the research and application of assembled concrete technology in the industry, under the inspiration of relevant foreign advanced concepts, the fully assembled precast concrete shear wall technology closely integrated with the characteristics of China's industry will develop rapidly and continuously improve.

1.3 Development Direction of Precast Concrete Shear Wall in China

To adapt to the basic national conditions of a large population and limited land, and to address the housing supply crisis brought about by the accelerated urbanization process, shear wall structures suitable for residential buildings that align with the housing consumption concept of "no protruding columns or beams" have become the primary structural form for high-rise and super high-rise residences in China. Therefore, the application object of prefabricated concrete shear wall technology in China will also be high-rise and super high-rise residential buildings. Considering that shear wall structures in foreign countries are generally used in low-rise or multi-story buildings, high-rise and super high-rise prefabricated concrete shear wall technology has become a major feature of prefabricated building technology in China, and the research and development of prefabricated concrete shear wall technology that meets the seismic requirements of high-rise and super high-rise building structures has become the macro development direction of this technology field in China.

Based on the research results of precast d concrete shear wall technology, it is suggested to carry out technical improvement and breakthrough in the next stage from two aspects: on the one hand, the current mainstream assembled monolithic precast concrete shear wall technology should be further optimized and improved, such as strengthening the grouting quality control and detection of steel grout sleeve connection technology and the applicability verification of steel grout anchored overlap connection. On the other hand, we should boldly break through the concept of "equivalent cast-in-place" and actively develop efficient fully assembled precast concrete shear wall technology, such as making full use of efficient dry connection technologies such as equivalent cast-in-place and bolted connection to avoid on-site wet operations. "Two-pronged" to promote the continuous development and improvement of China's precast concrete shear wall technology.

In summary, the development of prefabricated concrete shear wall in China should be oriented to the actual social needs of our country and the general demands of the industry. Based on the existing research foundation and international advanced practices, technical improvement and technological breakthrough should be carried out to form a prefabricated concrete shear wall technology with Chinese characteristics that is compatible with the development stage and industry level of prefabricated concrete technology in China.

References

1. Yee AA. Social and environmental benefits of precast concrete technology. PCI J. 2001;46(3):14–9.
2. Yee AA. Structural and economic benefits of precast/prestressed concrete construction. PCI J. 2001;46(4):34–43.
3. Several opinions of the CPC central committee and the state council on further strengthening the management of urban planning and construction [EB/OL] (2016). http://www.gov.cn/zhengce/2016-02/21/content_5044367.htm
4. Chinese Architecture Society. Construction industry modernization development outline defines industrialization development goals for 5 to 10 Years [EB/OL] (2017). http://www.chinaasc.org/news/116689.html
5. American Concrete Institute (ACI). ACI 550.1R-09 guide to emulating cast-in-place detailing for seismic design of precast concrete structures. Farmington Hills, MI. 2009.
6. Priestley M. Overview of PRESSS research program. PCI J. 1991;36(4):50–7.
7. Nigel Priestley M, Sritharan S, Conley J, et al. Preliminary results and conclusions from the PRESSS five-story precast concrete test building. PCI J. 1999;44(6):42–67.
8. Soudki KA, Rizkalla SH, Daikiw RW. Horizontal connections for precast concrete shear walls subjected to cyclic deformations Part 2: prestressed connections. PCI J. 1995;40(5):82–96.
9. Perez FJ, Pessiki S, Sause R. Lateral load behavior of unbonded post-tensioned precast concrete walls with vertical joints. PCI J. 2004; 48–64.
10. Perez FJ, Pessiki S, Sause R. Seismic design of unbonded post-tensioned precast concrete walls with vertical joint connectors. PCI J. 2004;49(1):58–79.
11. Perez FJ, Sause R, Pessiki S. Analytical and experimental lateral load behavior of unbonded posttensioned precast concrete walls. J Struc Eng. 2007;133(11):58–79.
12. ACI Innovation Task Group 5. ACI ITG 5.1–07 Acceptance criteria for special unbonded post-tensioned precast structural walls based on validation testing. Farmington Hills, MI: American Concrete Institute; 2007.
13. ACI Innovation Task Group 5. ACI ITG 5.2–09 Requirements for design of a special unbonded post-tensioned precast shear wall satisfying ACI ITG-5.1. Farmington Hills, MI: American Concrete Institute; 2009.
14. Kurama Y. Seismic design of unbonded post tensioned precast walls with supplemental viscous damping. ACI Struct J. 2000;97(4):648–58.
15. Kurama Y. Simplified seismic design approach for friction damped unbonded post tensioned precast walls. ACI Struct J. 2001;98(5):705–16.
16. Ajrab J, Pekcan G, Mander J. Rocking wall-frame structures with supplemental tendon systems. J Struct Eng. 2004;130(6):895–903.
17. Perez F, Pessiki S, Sause R. Seismic design of unbonded post-tensioned precast concrete walls with vertical joint connectors. PCI J. 2004;49(1):58–79.
18. Perez F, Pessiki S, Sause R. Lateral load behavior of unbonded post-tensioned precast concrete walls with vertical joints. PCI J. 2004;49(2):48–64.
19. Hamid NH, Mander JB. Lateral seismic performance of multipanel precast hollowcore walls. J Struc Eng 2010;136(7): 795–04.
20. Kurama Y. Hybrid post-tensioned precast concrete walls for use in seismic regions. PCI J. 2002;47(5):36–59.
21. Smith BJ, Kurama YC, McGinnis MJ. Design and measured behavior of a hybrid precast concrete wall specimen for seismic regions. J Struc Eng. 2007;137(10):58–79.
22. Sritharan S et al. Precast concrete wall with end columns (PreWEC) for earthquake resistant design. Earthq Eng Struct Dyn, 2015;44(12): 2075–92.
23. Lu X, Wu H, Zhou Y. Seismic collapse assessment of self-centering hybrid precast walls and conventional reinforced concrete walls. Struct Concr. 2017;18(6):938–49.
24. Gu A, Zhou Y, Xiao Y, et al. Experimental study and parameter analysis on the seismic performance of self-centering hybrid reinforced concrete shear walls. Soil Dyn Earthq Eng. 2019;116:409–20.

25. PEIKKO Oy. Tests on Peikko wall connections. Finland: Peikko Group Corporation; 2009. p. 1–7.
26. Bora C, Oliva MG, Nakaki SD, Becker R. Development of a precast concrete shear-wall system requiring special code acceptance. PCI J 2007; 2–15.
27. El Semelawy M, El Damatty A, Soliman AM. Novel anchor-jointed precast shear wall: testing and validation. Struct Build, 2015;168(4): 263–74.
28. Xue WC, Gu XL, Hu XX, et al. Experimental study on assembled monolithic concrete shear wall with bolted connection under low cyclic loading. J Civ Eng. 2014;47(S2):221–6.
29. Xue WC, Chu MX, Liu YN, et al. Seismic performance of new precast concrete shear wall under high axial compression ratio. J Harbin Eng Univ. 2018;37(3):452–60.
30. Dusicka P, Kay T. In-plane lateral cyclic behavior of insulated concrete form grid walls. J Struc Eng. 2011;137(10):1075–84.
31. Holden T, Restrepo J, Mander JB. Seismic performance of precast reinforced and prestressed concrete walls. J Struc Eng. 2003;129(3):286–96.
32. Industry standard of the People's Republic of China. Technical specification for precast concrete structures JGJ 1–2014. Beijing: China Architecture & Building Press; 2014.
33. National standard of the People's Republic of China. Technical standard for assembled buildings with concrete structure GB/T 51231–2016. Beijing: China Architecture & Building Press; 2016.
34. Chen X, Liu M, Biondini F, et al. Structural behavior of precast reinforced concrete shear walls with large-diameter bars. ACI Struct J. 2019;116(5):77–86.
35. Li J, Fan Q, Lu Z, et al. Experimental study on seismic performance of T-shaped partly precast reinforced concrete shear wall with grouting sleeves. Struct Design Tall Spec Build. 2019;28(13): e1632.
36. Wang D, Lv XL, Lu WS. Experimental study on seismic behavior of precast concrete shear walls with jointed connection beams. 2013;34(10):1–11
37. Jiao AL, Zhang P, Li YH, et al. Experimental study on seismic behavior of precast shear wall with ring-stiffened and anchored connection. 2015;36(5):103–9
38. Jiao AL, Zhang ZS, Gao YF, et al. Full-scale pseudo-dynamic experimental study on assembled concrete shear wall with loop bars and bolts. Industrial buildings. 2017;47(4):45–50.
39. Wang W, Li A, Wang X. Seismic performance of precast concrete shear wall structure with improved assembly horizontal wall connections. Bull Earthq Eng. 2018;16(9):4133–58.
40. Chu M, Liu J, Sun Z. Experimental study on mechanical behaviors of new shear walls built with precast concrete hollow moulds. Eur J Environ Civ Eng. 2019;23(12):1424–43.
41. Han W, Zhao Z, Qian J, et al. Experimental seismic behavior of squat shear walls with precast concrete hollow moulds. Earthq Eng Eng Vib. 2019;18(4):871–86.
42. Lian X, Ye XG, Wang DC, et al. Experimental study on seismic behavior of composite shear wall. J Hefei Univ Technol (Nat Sci Ed). 2009;32(8):105–9.
43. Xiao QD. Experimental and theoretical study on seismic performance of assembled concrete double-slab shear wall. Nanjing: Southeast University; 2015.
44. Zhang HM, Lv XL, Duan YF, et al. Nonlinear study of partially precast reinforced concrete composite walls (PPRC-CW). J Civ Eng. 2010;43(S2):93–100.
45. Shen SD, Pan P, Miao QS, et al. Test and analysis of reinforced concrete (RC) precast shear wall assembled using steel shear key (SSK). Earthquake Eng Struct Dynam. 2019;48(14):1595–612.
46. Sun J, Qiu HX, Jiang HB. Analysis of bearing capacity of bolted assembled reinforced concrete shear wall. 2019;8:23–30

Chapter 2
Emulative Hybrid Assembled Precast Concrete Shear Wall

2.1 Current Issues of Assembled Monolithic Precast Concrete Shear Walls in China

As mentioned earlier, the assembled monolithic precast concrete shear walls that have received significant research attention and application in China primarily utilize three forms of connections: grout sleeve connections, grout anchored overlap connections, and constrained grout anchored overlap connections. These have been recognized as having "equivalent cast-in-place" performance. Therefore, they have been formally incorporated into Chinese standards, such as "*Technical Specification for prefabricated concrete structures*" *(JGJ 1-2014)* [1] and "*Technical standard for assembled buildings with concrete structure*" *(GB/T 51231-2016)* [2].

Based on a summary analysis of the experimental results related to these three connection technologies for assembled monolithic precast concrete shear walls, the following three issues have been identified, which indicate that their performance still falls short in some aspects compared to traditional cast-in-place structures:

(1) Early cracking

As shown in Table 2.1, the cracking load of the assembled monolithic precast concrete shear wall specimen is between 75 and 100% of that of the reference cast-in-place concrete specimen.

The smaller cracking load means reduced safety of the structure or component under normal service conditions, and premature cracking will directly affect the normal service function and durability of the structure. It is considered that, due to the discontinuity of concrete at the joint of the assembled monolithic precast concrete shear wall, the contribution of concrete crack resistance in the calculation formula of the cracking load (see Formula 7.2.3-6 of *Code for design of concrete structures (GB 50010–2010)* [3]) is zero, which inevitably causes the cracking load to be lower than that of the cast-in-place shear wall. Meanwhile, the natural discontinuity of concrete also affects the initial or elastic stiffness of the structure or component. The low

Z. Zhu et al., *Research on Emulative Hybrid Assembled Precast Concrete Shear Walls*,
https://doi.org/10.1007/978-981-96-9578-2_2

Table 2.1 Statistics of partial test data in China

Data sources	Specimen number	Cracking load/kN	Cracking load ratio (%)	Specimen condition
Literature [4]	SW1	190	100	Cast-in-place specimen
	SW2	–	–	Prefabricated wall panel and cast-in-place boundary elements
	SW3	190	100	Fully precast, grout sleeve splicing of vertical rebars of boundary elements, disconnected vertical distribution rebars
	SW4	170	89	Fully precast, grout sleeve splicing of vertical rebars of boundary elements, vertical distribution rebars lapping in grout-filled hole
	SW5	0 (Cracking occurring after axial compression)	0	Partially precast, cast-in-place at the bottom of 250 mm height range, steel hoop connections for vertical rebars
Literature [5]	XJ	210	100	Cast-in-place specimen with horizontal construction joints
	JW	170	81	Fully precast external wall specimen, the joint between precast components moved 600 mm from the top of the base, vertical rebars lapping in grout-filled hole formed by metal corrugated pipe
	JN	210	100	Fully precast internal wall specimen, the joint between precast components located on the top of the base, vertical rebars lapping in grout-filled hole formed by metal corrugated pipe
Literature [6]	XJ	210	100	Cast-in-place specimen
	UN1	200	95	Partially cast-in-place boundary elements, U-shaped hoop connection for vertical rebars within cast-in-place concrete, vertical distribution rebars lapping in grout-filled hole formed by metal corrugated pipe
Literature [7]	PW1	150	100	Cast-in-place specimen
	PW2	115	77	Overlapping U-shaped vertical rebars grouted in hole formed by mesh steel plate

(continued)

Table 2.1 (continued)

Data sources	Specimen number	Cracking load/ kN	Cracking load ratio (%)	Specimen condition
	PW3	130	87	Overlapping U-shaped vertical rebars anchored in fine aggregate concrete casted in hole formed by mesh steel plate
	PW4	120	80	Overlapping heading vertical rebars grouted in hole formed by mesh steel plate
	PW5	120	80	Same as PW4
	PW6	120	80	Overlapping heading vertical rebars anchored in fine aggregate concrete casted in hole formed by mesh steel plate
	PW7	120	80	Same as PW6
Literature [8]	SW1	150	100	Cast-in-place specimen
	SW2	115	77	Boundary elements constrained by spiral stirrups, vertical rebars lapping in grout-filled hole formed by metal corrugated pipe
	SW3	130	87	Same as SW2

initial stiffness will directly lead to the large elastic displacement of the structure or component, accelerate the stiffness degradation, affect the comfort of the structure and cause other adverse consequences.

(2) Relatively concentrated concrete damage

As illustrated in Fig. 2.1, the final failure of assembled monolithic precast concrete shear wall is relatively concentrated in the horizontal gap between precast components. On the one hand, local damage may be beneficial to the repairing after the earthquake. On the other hand, concrete stress at the ends of the wall is more concentrated than that of the cast-in-place concrete. If the concrete is not well confined, early crushing may occur, thus affecting the bearing capacity and deformation capacity of components. Meanwhile, the concentrated concrete failure not only restricts the expansion of plastic hinge, but also affects the energy dissipation capacity of members to a certain extent, which is an important reason for the relatively low energy consumption capacity of the tested assembled monolithic precast concrete shear wall specimens.

(3) Limitation of force transmission characteristics of joints

At present, there are three popular forms of rebar connections: grout sleeve connections, grout anchored overlap connections, and constrained grout anchored overlap connections. These connections can be categorized into direct connection and indirect connection according to their force transmission characteristics. As for the internal

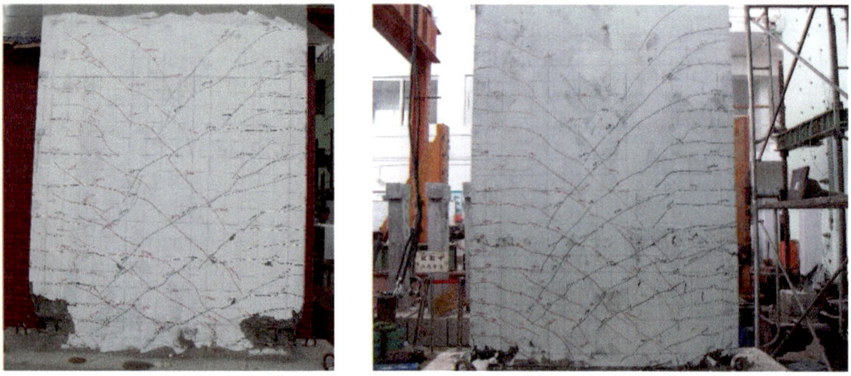

(a) Cast-in-place shear wall (b) Assembled monolithic precast concrete shear wall

Fig. 2.1 Comparison of failure modes of shear wall specimens

force of the rebar, the grout sleeve connection belongs to the direct force transfer connection, where the rebars only bear the axial force, which has no difference with the continuous rebars in the cast-in-place concrete structure. Therefore, the grout sleeve connection is considered to be the most reliable option. While the other two belong to indirect connection, rebars and surrounding concrete have to bear the adverse effects arising from eccentric force transmission, which is relatively unreliable. However, for the grout sleeve connection, due to its technological manufacturing problems, the sleeve section is larger than the rebar section, causing section non-uniformity that results in the sudden change of bearing capacity, stiffness, and potential weak section. As for the construction, due to the sudden change of the rebar and sleeve sections, the arrangement of stirrups around longitudinal rebars becomes a difficult problem. If the external size of stirrups remain unchanged, it will not be able to adapt to the sudden change of this section. If the external size changes, the thickness of the concrete cover outside the sleeve will be increased, and the concrete core area will be reduced, which is bound to affect the bearing capacity and ductility of the components. For indirect connection, the available experimental results show that the adverse effects caused by eccentric force transmission can be overcome through reasonable and reliable confinement measures without any manufacturing constraint.

Due to the above-mentioned problems in the performance of the assembled monolithic precast concrete shear wall, although it is considered to be equivalent to the cast-in-place concrete shear wall, its application is still limited. It can be seen from Table 1.1 that the maximum applicable height of the building is at least 10 m lower than that of the cast-in-place concrete shear wall structure under the same conditions. Nowadays this situation does not obviously affect its engineering application. However, in the long run, it will cause the industry and society to misjudge that its seismic performance is inferior to that of cast-in-place concrete shear wall, which will eventually be detrimental to the competition with the cast-in-place concrete shear wall structural technology in scientific market.

2.2 Proposal of Emulative Hybrid Precast Concrete Shear Wall

Based on the performance issues of the assembled monolithic precast concrete shear wall, in order to completely break through its current application limitations, a hybrid precast concrete shear wall (Emulative Hybrid Shear Wall System, hereinafter referred to as EHSW technology) is proposed by incorporating international advanced practice and current domestic research achievements. The schematic diagram of EHSW structure is shown in Fig. 2.2.

In EHSW technology, vertical reinforcement still employs metal corrugated tube grout anchored overlap connection technology, primarily considering the following factors related to production, construction, and cost:

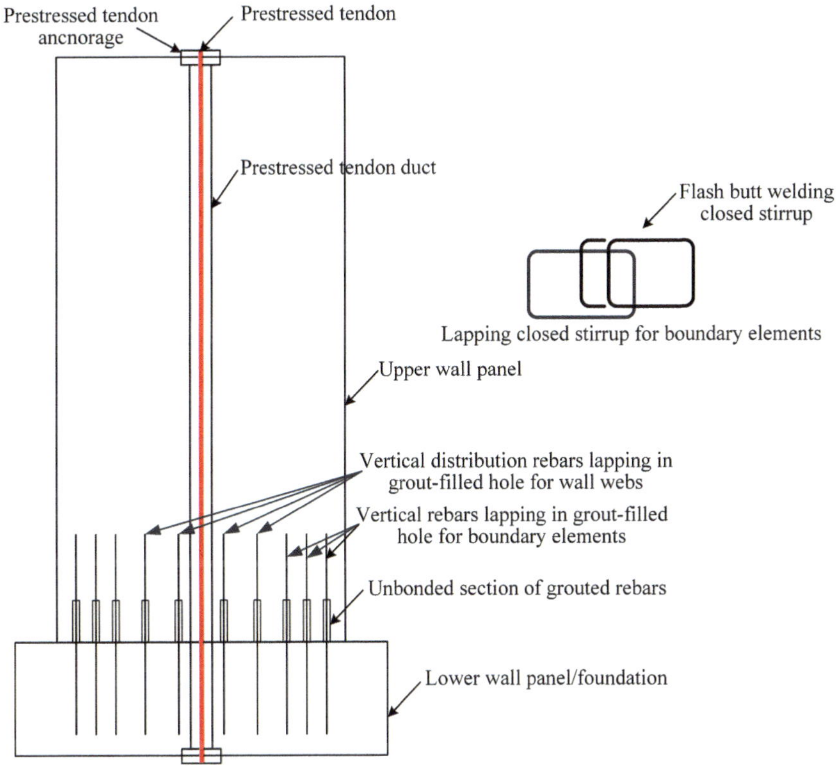

Fig. 2.2 Schematic diagram of EHSW

(1) Constrained lapping in grout-filled hole adopts core-pulling to form the hole, which is difficult in process and the quality is not easy to ensure, while metal corrugated bellows are embedded to make the grouted holes, which is convenient for process and reliable in quality.
(2) Due to the constraint of sleeve specification, the gap between rebars and sleeve inner wall is relatively small, so the adjustable tolerance is also exceedingly small. To ensure the smooth on-site docking of precast components, the manufacturing accuracy is relatively stringent, and the error is generally required to be controlled at the millimeter level, which is a big challenge for the current component prefabrication technology in China.
(3) In terms of cost, the cost of sleeve is the highest, which is generally several times higher than that of traditional rebar connection. Driven by objective of reducing cost and pursuing profit, enterprises are not motivated to adopt the grout sleeve independently. Although the use of metal corrugated bellows will bring extra cost increase, it can generally be controlled around the cost of traditional rebar connection.

The EHSW technology, while continuing to use the metal corrugated tube grout anchored overlap connection technology, integrates the following techniques to address the above-mentioned three main issues:

(1) Post-tensioned unbonded prestressing technique. Different from the concept of foreign hybrid precast concrete shear wall systems or unbonded pre-stressed precast concrete shear wall systems that use prestressed compression as the main measure to connect precast shear wall panels, EHSW introduces prestressed compression technology. The main purpose is to solve the problem of early cracking of the fabricated concrete shear wall and lower initial stiffness. At the same time, it is expected that the restoring force provided by the unbonded prestressed tendons can reduce the residual deformation of the component after the earthquake. Therefore, the arrangement of prestressed tendons is different from that of foreign technologies. Unbonded prestressed tendons of EHSW are only arranged in the middle of the wall segments, rather than being distributed along the wall segments, and they are symmetrically aligned along the centerline of the wall segments. Based on this, it should be possible to properly solve the problem of earlier cracking of the monolithic precast concrete shear wall.
(2) Local unbonded technology of grouted rebars. In foreign countries, the grouted rebars crossing the horizontal joint are also dealt with local unbonded treatment, mainly to keep the rebar elastic and limit the shear slip of precast components along the horizontal joint. Local unbonded technology can also avoid low cycle fatigue failure of rebars under high strain state. As for EHSW, in addition to the foregoing purposes, it is more important to expect tensile stress transferred to the concrete far away from the horizontal joint, so as to improve the stress state of the concrete near the joint. By manually transferring or extending the possible destruction of concrete area, the above-mentioned problem of relatively concentrated damage of precast concrete shear wall can be improved.

(3) Lapping closed stirrup technology. The available test results have fully proved the effectiveness of lapping closed stirrup in restraining concrete and improving the overall seismic performance of the components. EHSW adopts lapping closed stirrup technology within the range of the rebar lapping in grout-filled hole of its boundary elements, which on the one hand is to strengthen the restraint performance of concrete, on the other hand, for the indirect connection, is to improve the force transmission performance, so as to overcome the aforementioned problem in force transmission of connection structure.

2.3 Key Parameters of EHSW

According to the structural characteristics of EHSW, the key parameters affecting its seismic performance include the section area of unbonded prestressed tendon, initial effective stress of unbonded prestressed tendon, local unbonded length of grouted rebar, axial compression ratio, etc. The key parameters are analyzed as follows.

(1) Section area of unbonded prestressed tendon

Due to the introduction of unbonded prestressed connection technology, the analysis of EHSW seismic mechanism should take into account the important seismic design parameters of prestressed concrete structures, that is, the prestress strength ratio and the height of concrete compression zone, which are important indexes affecting the seismic performance of prestressed concrete structures and the related seismic design process. The prestress strength ratio can be calculated according to Formula 2.1, quoted from Article 3.1.9 of *Standard for seismic design of prestressed concrete structures (JGJ/T 140–2019)* [9]).

The preliminary analysis shows that, under the same other conditions, the larger section area of prestressed tendon, the higher prestress strength ratio, and the higher concrete compression zone, which leads to more unreliable earthquake resistance for EHSW. *Standard for seismic design of prestressed concrete structures (JGJ/T 140–2019)* has made the upper limit on the prestressed strength ratio or/and the height of concrete compression zone for various structural components (including frame beams, frame columns, slabs, etc.) in different seismic grades. Although there are no explicit or clear provisions for shear walls in *Standard for seismic design of prestressed concrete structures (JGJ/T 140–2019)*, EHSW can still be treated as frame beams based on the fact that EHSW is limited to bending failure or bending-shear failure. Therefore, the section area of prestressed tendon should be controlled to meet the requirement of the prestressed strength ratio. At the same time, a favorable aspect of EHSW is that the sleeve-anchor reinforcement can still improve the stress in the component section and its energy-dissipating function, thereby significantly enhancing its seismic performance.

$$\lambda = \frac{\sigma_{pu} A_p h_p}{\sigma_{pu} A_p h_p + f_y A_s h_s} \tag{2.1}$$

λ prestressed strength ratio,
σ_{pu} design stress value of unbonded prestressed tendon,
f_y design value of tensile strength of ordinary reinforcement,
A_p section area of prestressed rebar in tensile zone,
A_s area of ordinary rebar in tensile zone,
h_p effective distance between the resultant point of longitudinal tensile prestressed rebar and the compression edge of the section,
h_s effective distance between the resultant point of longitudinal tensile ordinary rebar and the compression edge of the section.

(2) Initial effective stress of unbonded prestressed tendon

Under the same other conditions, the higher the initial stress of unbonded prestressed reinforcement, the stronger the resistance of EHSW to cracking. In addition, according to formula 2.2 (quoted from Article 5.1.11 of *Technical specification for concrete structures prestressed with unbonded tendons (JGJ 92–2016)* [10]), it can be found that the initial stress of unbonded prestressed tendon is also an important parameter to determine its final stress value, thus determining the bearing capacity of EHSW.

$$\sigma_{pu} = \sigma_{pe} + \Delta\sigma_p \tag{2.2}$$

σ_{pu} design stress value of unbonded prestressed tendon,
σ_{pe} initial effective stress of unbonded prestressed tendon,
$\Delta\sigma_p$ stress increment in unbonded prestressed tendon.

(3) Local unbonded length of grouted rebar

The preliminary analysis shows that, the absence of local bonding of the grouted rebar can avoid the concentration of rebar strain at the section of horizontal joint, reduce the rebar strain, delay the yield of the rebar, but increase the yield deformation. Meanwhile, the larger the unbonded length, the greater the usable deformation of the steel bar after yielding. Under the premise of no premature collapse of concrete, the ultimate deformation capacity of components can be improved. Besides, he length of unbonded concrete is also related to the transfer of the plastic failure zone of concrete, and directly determines the extended length of grouted rebars from the perspective of cost.

(4) Axial compression ratio

Axial compression ratio is an important control index for shear wall structure design, especially for ductile seismic design. Similar to compression-bending components, for shear walls controlled by bending failure, the fundamental factors affecting their ductility are still the height of the compression zone and the ultimate compressive strain of the concrete. Both the decrease of compression zone height and the increase of concrete ultimate compressive strain can increase the ultimate curvature of section and improve ductility, and vice versa. In view of this, generally speaking, when the axial compression ratio increases, the section bearing capacity will be improved, but this will lead to the increase of the compression zone height, resulting in a significant reduction in ductility. Therefore, current national standard provisions such as Article 11.7.16 of *Code for design of concrete structures (GB50010-2010)* [3], Article 7.2.13 of *Technical specification for concrete structures of tall building (JGJ3-2010)* [11] and Article 6.4.2 of *Code for seismic design of buildings (GB50011-2010)* [12] all put forward clear control requirements for axial compression ratio of shear walls. To control the axial compression ratio of the shear wall is to control the vertical load and the normal stress of the section of the shear wall. It can reduce the height of the compression area of the section concrete, make the plastic hinge of the shear wall have sufficient rotation capacity at the bottom strengthening zone, and enhance the ductility of components and structures.

2.4 Detailed Configuration of Hybrid Precast Concrete Shear Wall

Based on the comprehensive adoption of unbonded post-tensioned prestressing technology, local unbonded technology of grouted rebars, and lapping closed stirrup technology in EHSW, in order to achieve the relevant technical objectives and facilitate the actual engineering construction, and taking into account the needs of subsequent test specimen manufacture, proper design needs to conducted for the anchoring and tensioning device of unbonded prestressed tendons, the construction of unbonded grouted rebars and lapping closed stirrups. The specifications are detailed below:

(1) Anchorage and tensioning devices of unbonded prestressed tendon

As shown in the schematic diagram of EHSW (Fig. 2.2), the unbonded prestressed tendon is arranged through the whole wall height, that is, the unbonded prestressed tendon should be continuous in the upper and lower wall panels and the tensile stress should also be consistent. Considering the actual construction conditions of engineering project, EHSW should be prefabricated, installed, and tensioned by layers. Therefore, it is necessary to design suitable anchoring and tensioning devices of unbonded prestressed tendons for reliable anchorage and convenient tensioning.

Prestressed steel strand and prestressed threaded steel bar are considered to be used for unbonded prestressed tendon of EHSW. As for the prestressed steel strand, the anchor ring pressure-bearing anchorage is made by updating the clamp anchorage, as shown in Fig. 2.3a. By increasing the thickness of the original clamp anchorage, processing the screw teeth around the anchorage, and covering the anchor ring, after the steel strand is tensioned in place, the anchor ring is tightened on the concrete surface to form a reliable anchorage of unbonded prestressed tendons. The device can realize repeated tension and tightening, and further reduce the prestress loss caused by clip retraction, anchorage compression and screw clearance. The front anchor ring of the device can be anchored, while the peripheral screw of the back-end Anchorage and the back-end anchor ring which has the same structure as the front

(a) Prestressed steel strand

(b) prestressed threaded steel bar

Fig. 2.3 Unbonded pre-stressed reinforcement anchoring and tensioning device

anchor ring and only doubles in thickness can realize the lengthening of unbonded prestressed tendons. Meanwhile, it is also convenient for the test specimen to release the prestressed tendons and disassemble the anchorage, so as to recycle the anchorage to save the test cost as much as possible. As for prestressed threaded steel bar, its products are equipped with special anchors and extenders, as shown in Fig. 2.3b. Therefore, the anchorage and extender can be directly used to realize the anchoring and continuous requirements of unbonded prestressed tendons, and it is only necessary to ensure the sufficient cutting length of prestressed threaded steel bars in the process of member design and construction.

(2) Local unbonded construction of grouted rebars

To achieve the unbonded condition of the grouted rebar near the horizontal joint and its surrounding grout mix, it is considered to use an outer PVC hose, sealed at both ends with tape, or to wind tape along the entire unbonded length. The former process is relatively complex, but the finished product is well protected, and the PVC hose is not easy to be scratched by coarse aggregate during concrete pouring, which affects the unbonded effect. The latter process is relatively simple, but it is not conducive to the protection of finished products. In the process of component transportation and concrete pouring, surface damage is easy to occur, which makes the grouting mixture pollute the unbonded section of steel bars.

Additionally, although preliminary performance tests for the anchorage of the joint indicated that a bonded length of 0.6lab for corrugated tube grouted rebar could meet the requirements, a more conservative approach generally applies a factor of 1.10 for safety. The length of EHSW grouted rebar will remain constant; therefore, its local unbonded structure will reduce its anchorage length, and its applicability and reliability can be further verified through specimen testing.

(3) The construction of lapping closed stirrups

In order to ensure the stress continuity of stirrups and improve their confinement performance on concrete, EHSW adopts welded closed stirrups instead of traditional closed stirrups with 135° lapping hooks. As shown in Fig. 2.4a, closed stirrups are

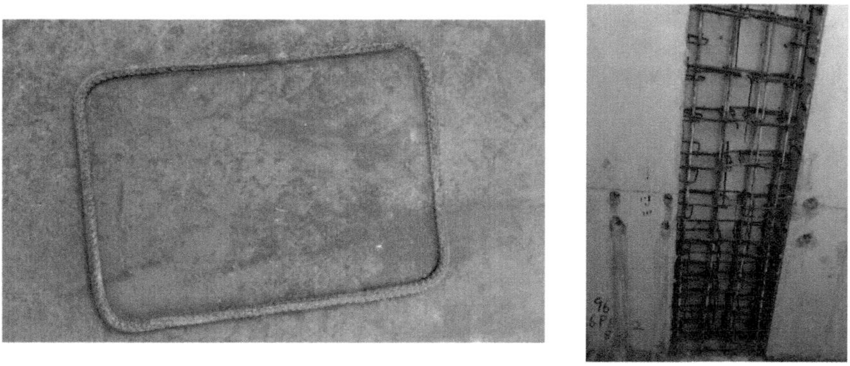

(a) Flash butt weld closed stirrup (b) Field application photo

Fig. 2.4 Lapping closed stirrup construction

made with single steel bar, and flash butt welding is adopted for joints. In order to further improve the confinement effect of closed stirrups on concrete, the traditional multiple stirrups with large and small stirrups and tie bars are replaced by the lapping closed stirrups, which reduces the ratio of length to diameter of stirrups, improves the constraint stiffness of stirrups and strengthens the hoop effect on core concrete. The lapping closed stirrups have been applied in relevant projects for the horizontal steel bar connection within the cast-in-place concrete between the adjacent precast wall panels, which can be seen in Fig. 2.4b.

References

1. Industry standard of the People's Republic of China. Technical specification for precast concrete structures JGJ 1–2014. Beijing: China Architecture & Building Press; 2014.
2. National standard of the People's Republic of China. Technical standard for assembled buildings with concrete structure GB/T 51231–2016. Beijing: China Architecture & Building Press; 2016.
3. National standard of the People's Republic of China. Code for design of concrete structures GB 50010–2010 (2015 Edition). Beijing: China Architecture & Building Press; 2010.
4. Qian JR, Yang XK, Qin H, et al. Seismic performance test of precast reinforced concrete shear wall with different connection methods of vertical reinforcement. J Build Struct. 2011;32(6):51–59.
5. Chen YG, Liu JB, Guo ZX, et al. Seismic performance test of assembled shear wall with horizontal seam steel bar grout anchor lap joint. J Harbin Inst Technol. 2013;45(6):83–89.
6. Liu JB, Chen YG, Guo ZX, et al. Experimental study on seismic performance of U-shaped closed bar connection in horizontal joints of assembled concrete shear wall. J Southeast Univ (Nat Sci Ed) 2013;43(3): 565–70.
7. Liu JB, Chen YG, Guo ZX, et al. Experimental study on seismic performance of assembled shear wall with new vertical connection. J Hunan Univ (Nat Sci Ed) 2013;41(4): 16–24.
8. Liu JB, Chen YG, Guo ZX, et al. Seismic performance of spiral stirrup confined corrugated pipe grout-anchored fabricated shear wall. J South China Univ Technol (Nat Sci Ed). 2014;42(1):92–98.

9. Industry standard of the People's Republic of China. Standard for seismic design of prestressed concrete structures JGJ 140–2019. Beijing: China Architecture & Building Press; 2019.
10. Industry standard of the People's Republic of China. Technical specification for concrete structures prestressed with unbonded tendons JGJ 92–2016. Beijing: China Architecture & Building Press; 2016.
11. Industry standard of the People's Republic of China. Technical specification for concrete structures of tall building JGJ 3–2010. Beijing: China Architecture & Building Press; 2010.
12. National standard of the People's Republic of China. Code for seismic design of buildings GB 50011–2010 (2016 Edition). Beijing: China Architecture & Building Press; 2016.

Chapter 3
Experimental Study on Seismic Performance of Emulative Hybrid Assembled Precast Concrete Shear Wall

3.1 Test Purpose

The EHSW technology, which integrates post-tensioned unbonded prestressing technology, local unbonded technology of grout-anchored steel bars and closed stirrup technology, can reasonably solve the key problems existing in the current "equivalent cast-in-place" prefabricated concrete structure. In order to verify the feasibility and theoretical correctness of EHSW technology, it is necessary to carry out relevant experimental research. On the one hand, through the production of test specimens, the related construction process problems can be solved; on the other hand, it is also more critical. The actual seismic capacity of EHSW specimens can be mastered through experiments, and their seismic performance can be truly evaluated to prove the rationality and reliability of EHSW technology.

Aiming at the most basic anti-seismic component of EHSW technology, namely straight-shaped shear wall, a series of straight-shaped EHSW specimens are made and the anti-seismic performance test is carried out. The following research purposes are mainly achieved through the test:

(1) To investigate the crack development process, failure modes, strength, stiffness, deformation capacity, ductility, and energy dissipation capacity of the EHSW specimens under horizontal low-cycle repeated loading, and to compare these with cast-in-place control specimens to evaluate their seismic capacity.
(2) To explore the influence of key parameters related to EHSW technology on the seismic performance indicators of bearing capacity, stiffness, ductility, energy dissipation capacity, and residual deformation, providing data for establishing the design methodology.
(3) To understand the hysteresis characteristics of the EHSW specimens, analyzing their hysteresis curves and skeleton curves to provide a data foundation for establishing the restoring force characteristics of EHSW specimens.

Z. Zhu et al., *Research on Emulative Hybrid Assembled Precast Concrete Shear Walls*,
https://doi.org/10.1007/978-981-96-9578-2_3

3.2 Test Scheme

A series of one-shaped EHSW specimens and cast-in-place contrast specimens considering the parameters such as the cross-sectional area of unbonded prestressing tendons, the pretension of unbonded prestressing tendons, the local unbonded length of grout-anchored tendons and the axial compression ratio were fabricated. The seismic performance of the specimens was evaluated by horizontal low-cycle repeated loading test, and the influence of each parameter on the seismic performance indexes such as bearing capacity, stiffness, ductility, energy dissipation capacity and residual deformation of the specimens was analyzed.

3.2.1 Specimen Design

(1) Specimen size

In order to directly reflect the seismic performance of the components, simulate the actual construction process, and take into account the particularity of the assembled concrete specimens to ensure the realization of the connection structure, the test specimens were designed and manufactured in a 1:1 full-scale ratio.

According to Article 11.7.12 of the "*Code for design of concrete structures*" (*GB50010-2010*) [1], the wall limb cross-sectional thickness of shear walls should not be less than 160 mm for seismic levels I and II, and it should not be less than 1/20 of the story height or clear span length; for seismic levels III and IV, it should not be less than 140 mm and should not be less than 1/25 of the story height or clear span length. For the bottom reinforcement of shear walls at seismic levels I and II, the thickness should not be less than 200 mm and should not be less than 1/16 of the story height or clear span length. When there are no end columns or wing walls at the wall ends, the wall thickness should not be less than 1/12 of the story height or clear span length." In this experiment, since it is designed according to seismic level I, the thickness of the wall limb (wall thickness) is set at 200 mm.

"*Technical specification for concrete structures of tall buildings*" (*JGJ 3-2010*) [2] Article 7.1.8 stipulates that "short-pier shear wall refers to the shear wall whose section thickness is not greater than 300 mm and whose maximum ratio of section height to thickness is greater than 4 but not greater than 8". Conversely, for general shear walls with a thickness of less than 300 mm, the ratio of the wall limb height (wall length) to thickness (wall thickness) should be greater than 8. The wall length of the test specimen is 1700 mm, and the calculated limb thickness ratio is 8.5, which meets the requirements of shear wall section size.

In practical engineering structures, the mechanical behavior of shear wall is similar to that of cantilever beam. Due to the high height of shear wall, the failure mode of the bottom reinforcement part is generally controlled by bending failure or bending shear failure. Therefore, the test specimen wall height of 3300 mm, coupled with the top of the loading beam (beam height of 320 mm, loading point is located in the center of the beam), the considered wall root (seam) section of the shear span ratio of 3460/1700 ≈ 2.0, to ensure the formation of bending failure mechanism, to avoid shear failure or slip failure and other adverse phenomena.

Therefore, the basic size of the wall part of the test specimen is 200 mm (wall thickness) × 1700 mm (wall length) × 3300 mm (wall height). A reinforced concrete base is set at the bottom of the specimen to simulate the base plate or the lower shear wall in the actual project. In the test, the size design of the base should also consider the connection with the laboratory ground, the installation height of the loading equipment on the laboratory reaction wall and the bearing capacity requirements. The basic size of the test specimen base is 2200 mm (along the wall length direction) × 700 mm (along the wall thickness direction) × 640 mm (along the wall height direction). At the same time, a reinforced concrete top beam is set at the top of the specimen as a loading beam to connect the specimen with the loading equipment. The size design of the loading beam mainly considers the installation height of the loading equipment on the laboratory reaction wall, the size of the steel plate connected to the front end of the loading equipment and the loading shear span ratio of the section of the specimen wall root (abutted seam). The basic size of the loading beam in this test is 1700 mm (along the wall length direction) × 240 mm (along the wall thickness direction) × 320 mm (along the wall height direction).

The test specimen template is shown in Fig. 3.1.

(2) Specimen parameters

① Materials

Each part of the specimen (including the base, shear wall and loading beam) are used C35 strength grade concrete pouring; reinforcement of specimens, such as vertical steel bar, horizontal steel bar, stirrup of edge member and structural reinforcement of loading beam and base, all adopt HRB400 hot rolled steel bar. Unbonded prestressed tendons using 1860 Φ15.24 prestressed steel strand; the slurry anchor pipe adopts Φ40 finished metal bellows; the grouting material adopts BY (S) − 40 equipment foundation grouting material. The 1d strength of the product reaches more than 20 MPa, the 3d strength reaches more than 40 MPa, and the 28d strength can reach more than 60 MPa.

② Section area of unbonded prestressing tendons

Combined with the previous analysis results and Formula 2.1, the introduction of ordinary steel bars increases the denominator of Formula 2.1, thus reducing the prestressed strength ratio. In this test, the configuration of ordinary steel bars is kept unchanged, and the change of prestressed strength ratio is considered by changing the cross-sectional area of prestressed bars, so as to explore the influence of cross-sectional area parameters of prestressed bars on the seismic performance of specimens. At the same time, the unbonded prestressed tendons in this test are all 1860 Φ15.24 prestressed steel strands. Therefore, the change of the cross-sectional area of the prestressed tendons is reflected in the change of the number of prestressed tendons. At the same time, considering the limitation of the thickness of the wall

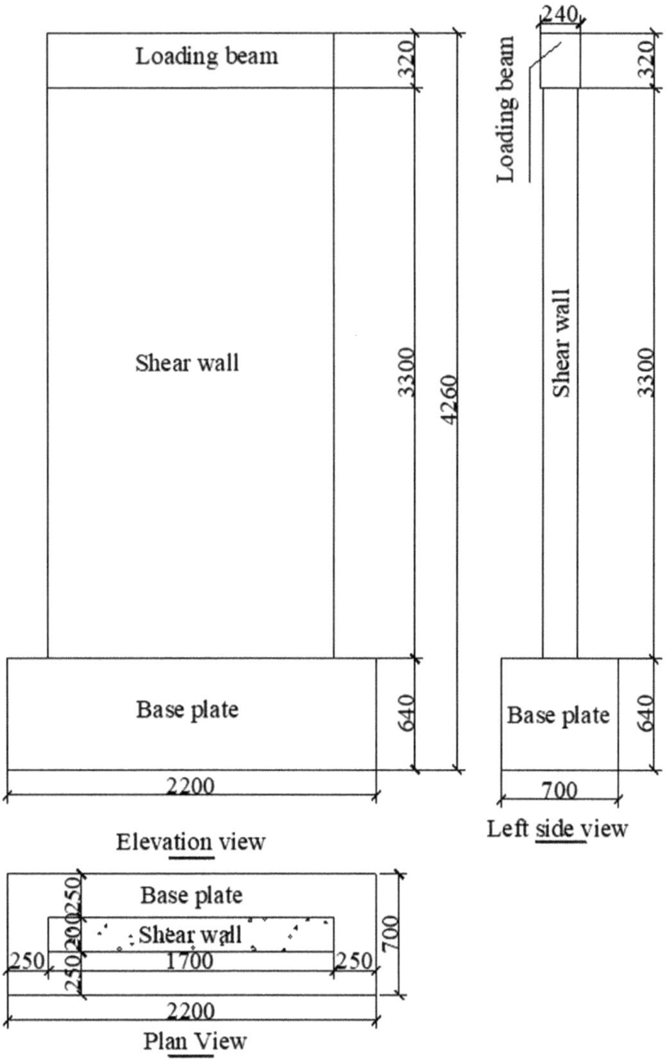

Fig. 3.1 Template of specimen

section on the size of the prestressed tendon anchorage, the number of prestressed tendons in three cases of 2, 3 and 4 is considered in this test, and the area is 280 mm^2, 420 mm^2 and 560 mm^2 respectively.

③ Initial effective stress of unbonded prestressed tendon

Combined with the above analysis results, with reference to the relevant engineering experience of unbonded prestressed concrete structures, the initial effective stress of three different prestressed tendons about $0.30f_{ptk}$, $0.40f_{ptk}$, $0.50f_{ptk}$ is considered in this test, and the corresponding initial stress theoretical values are 558, 744 and 930 MPa.

④ Local unbonded length of grout-anchored rebar

Combined with the previous analysis results, and from the specific situation of the specimen (mainly the thickness of the wall section), the local unbonded length of the slurry anchor steel bar is based on 200 mm and graded according to ± 50 mm. Therefore, this test specimen considers three unbonded lengths of 150, 200 and 250 mm.

⑤ Axial compression ratio

Combined with the previous analysis results, considering the actual specimen test, the determination of axial compression ratio still need to consider the ability of loading equipment. On the one hand, jacks commonly used to apply axial pressure have limited capacity (laboratory jacks are generally used within 250t, and large jacks are not easy to install because of their large size), and the laboratory ground bearing capacity is fixed (structural laboratory ground design bearing capacity is 50t per hole); on the other hand, the limit value of the axial compression ratio in the relevant standard provisions refers to the design axial compression ratio. In the calculation of the axial compression ratio, the load partial coefficient and the material strength partial coefficient are considered. There is a conversion relationship with the standard value of the material strength and the test value of the applied axial compression. The specific conversion is shown in the following. Here, only the design axial compression ratio considered in the preset test is given as 0.1 and 0.2 (the prototype of the test specimen is considered as the first-level (the seismic fortification intensity is 7° or 8°) shear wall, and the axial compression ratio limit is 0.5). The axial force of the test load will be calculated according to the test axial compression ratio and the measured value of the material performance converted from the design axial compression ratio.

(3) Specimen related structure

The details of the specimen design are shown in Fig. 3.2, and the relevant dimensions are measured in mm. The details of the structure are as follows.

① Structure of ordinary steel bar of specimen

The prototype of the test specimen is considered as a first-class (seismic fortification intensity of 7° or 8°) shear wall, and seismic construction measures are taken according to the relevant provisions of "*Code for seismic design of buildings*" (*GB 50011-2010*) [3].

The edge member type of the test prototype is the constraint edge member (dark column form). The length of the dark column and the reinforcement structure are safely used when the axial compression ratio is greater than 0.3. Therefore, according to Artical 6.4.5 of "*Code for seismic design of buildings*" (*GB 50011-2010*) [3], the length of hidden column is 400 mm > 0.20 h_w = 340 mm; the stirrup configuration of the hidden column is Φ8 @ 100 (HRB400). According to Formula 3.1, the calculated

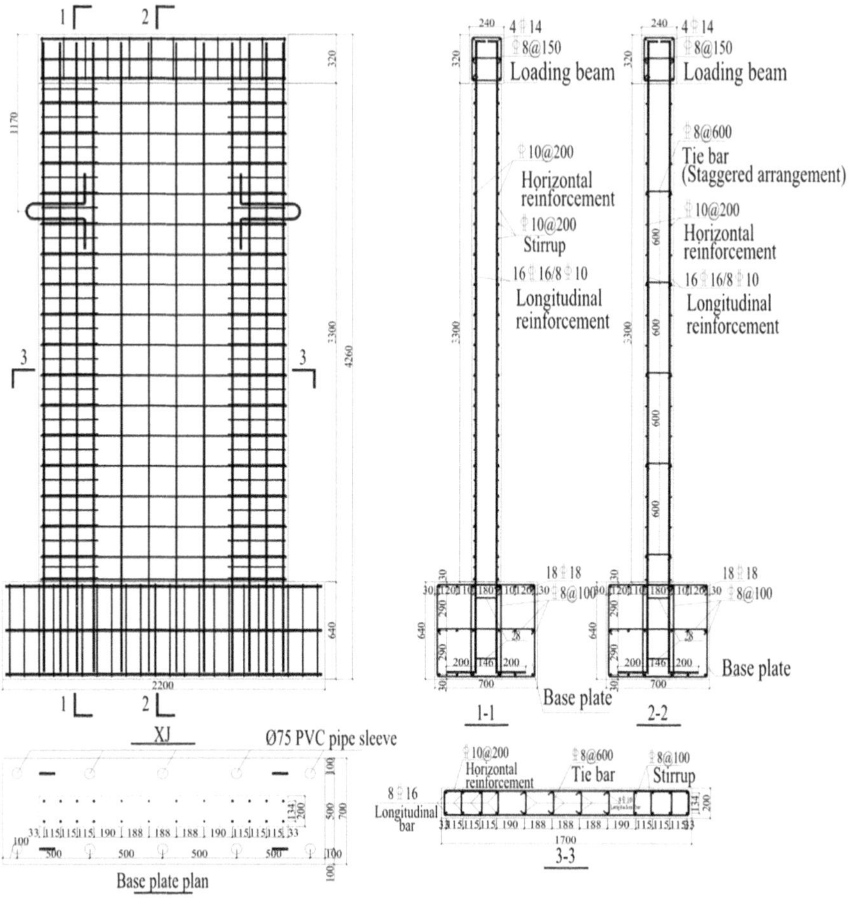

(a) Cast-in-place specimens

Fig. 3.2 Specimen design details

stirrup characteristic value is 0.22 > 0.20; the arrangement of longitudinal reinforcement is 8Φ16 = max $\{0.012\,A_c = 960\ \text{mm}^2, 8\Phi16\}$; stirrups or reinforcement along the vertical spacing of 100 mm.

$$\lambda_v = \frac{\rho_v f_{yv}}{f_c} \tag{3.1}$$

λ_v Characteristic value of stirrups in the hidden column area;
ρ_v Volumetric stirrup ratio in hidden column area, calculated as 1.01%;
f_{yv} Stirrups, tension reinforcement tensile strength design value, take 360 MPa;
f_c Design compressive strength of concrete under axial loading, taken as 16.7 MPa.

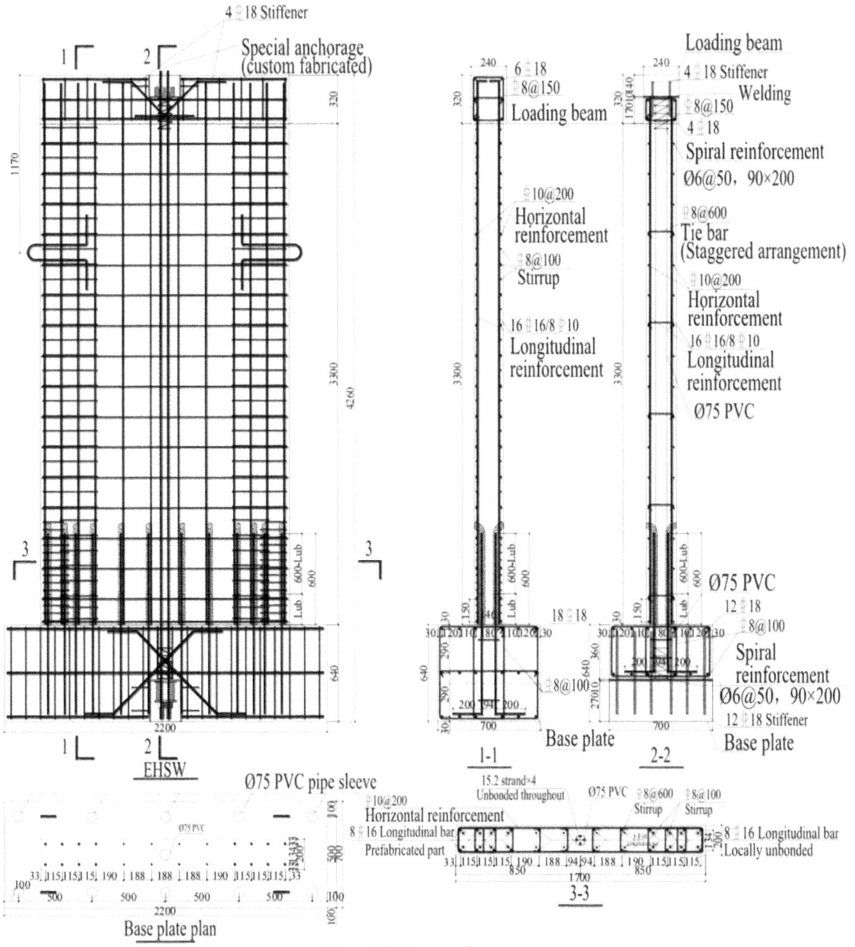

(b) EHSW specimen

Fig. 3.2 (continued)

According to the "*Code for seismic design of buildings*" (*GB 50011-2010*) [3] *6.4.3, 6.4.4*, the vertical and horizontal distribution of steel bars in shear walls are double-layer reinforcement. The vertical distribution of steel bars is 8Φ10 (HRB400), the spacing is about 188 mm, and the vertical distribution of steel bars is calculated. The reinforcement ratio is 0.42%, which meets the requirements of the minimum reinforcement ratio of the vertical distribution of steel bars in the first-level seismic wall is not less than 0.25%, the vertical reinforcement diameter is not more than 1/10 (20 mm) of the wall thickness and not less than 10 mm, and the spacing is not more than 300 mm; the horizontal distribution steel bar is configured as Φ10 @ 200 (HRB400), and the calculated horizontal distribution steel bar reinforcement ratio is 0.39%, which meets the requirements of the minimum reinforcement ratio of the horizontal distribution steel bar of the first-level seismic wall not less than 0.25%, the diameter of the horizontal steel bar not more than 1/10 (20 mm) of the wall thickness and not less than 8 mm, and the spacing not more than 300 mm; the tension reinforcement is configured as Φ8 @ 600 (HRB400) to meet the requirements that the spacing between the tension bars of the double-row distribution steel bars is not greater than 600 mm and the diameter is not less than 6 mm.

② Construction structure of unbonded prestressed tendon

The anchorages of unbonded prestressed tendons in EHSW specimens are all four-hole clip anchorages anchorage size is Φ100 × 40. Among them, the loading beam end is used as the tension end, and the base end is used as both the anchorage end and the pressure sensor installation part. To facilitate the installation of unbonded prestressed tendons, Φ75 PVC pipe is embedded in the center of loading beam, shear wall and base. At the same time, transverse through grooves are arranged at the top of the loading beam and the bottom of the base. The groove width of the loading beam is 220 mm and the depth is 150 mm, and the groove width of the base is 220 mm and the depth is 280 mm.

The anchorage and tension device of unbonded prestressed tendon adopts the special anchor ring pressure-bearing anchorage introduced in Sect. 2.4. The anchorage end makes the extrusion head for the steel strand, and the pressure sensor is pressed between the clip anchorage and the specimen. In addition, for the groove between the loading beam and the base, the cross oblique reinforcement is set up to strengthen it, and the spiral reinforcement is set up in the range of local confined concrete, so as to avoid the premature collapse of confined concrete or the cracking of members under local concentrated load.

The construction photos of unbonded prestressed tendons of EHSW specimen are shown in Fig. 3.3.

③ Pulp anchor steel structure

The metal bellows are pre-embedded close to the vertical reinforcement of the shear wall, and the two are tied with a tie wire; in the steel template side wall through the bolt positioning rubber plug, metal bellows into the rubber plug, in order to achieve accurate positioning of metal bellows; for all EHSW specimens, the specifications and length of the metal bellows are the same, the length of the straight section is 600 mm, and then the side of the shear wall is bent out through the arc to facilitate grouting. For the realization of the unbonded section, considering the convenient processing of the Specimen, the plastic film is wrapped in the unbonded section of the slurry anchor steel bar and wrapped and sealed with an electrical tape.

Construction method of grouted anchor steel bar is shown in Fig. 3.4.

④ Buckled closed stirrup structure

The edge component of the EHSW specimen adopts the buckled closed stirrups, and the diameter and spacing of the stirrups are consistent with the cast-in-place specimen, but the external stirrups of the four longitudinal reinforcements at the end

(a) Base PVC pipe embedded and local concrete reinforcement

(b) Wall body PVC pipe embedded and local pressure concrete strengthening

(c) Base steel strand extrusion anchor head and pressure sensor installation

(d) Loading beam end steel strand tension end equipment installation

Fig. 3.3 Construction photos of unbonded prestressed tendon of EHSW specimen

(a) Metal bellows installation

(b) Local unbonded approach

Fig. 3.4 Construction method of grouted anchor steel bar

Fig. 3.5 Construction method of buckled closed stirrup

are further encrypted, that is, the spacing is 50 mm, and the relevant photos are shown in Fig. 3.5.

(4) Specimen statistics

According to the set parameters, this test produced a total of 1 piece of cast-in-place shear wall specimens and 8 pieces of EHSW specimens, the key parameters of the relevant specimens are shown in Table 3.1.

3.2.2 Specimen Making

All parts of the cast-in-place specimen, including the base, shear wall, and loading beam, were cast in one operation. The EHSW specimens were divided into two pre-cast parts: the base and other components. The prefabricated wall is embedded with metal corrugated pipe, and the base is reserved with slurry anchor steel bar at the corresponding position. After the concrete reaches the design strength, the two parts are assembled, and the high strength grouting material is poured into the metal corrugated pipe. After the grouting material reaches the design strength, the specimen is transported to the laboratory and the unbonded prestressed tendons are pierced, tensioned and anchored.

EHSW specimen production process related photos are shown in Fig. 3.6.

3.2.3 Material Properties

The concrete is made of commercial concrete, and 150 mm square concrete test blocks are made at the same time of concrete pouring. A total of 3 groups are made, 6 test

Table 3.1 Statistical table of specimens

Numbering	Number of prestressing tendons	Initial prestress/MPa	Initial prestress force/kN	Unbonded length of grout anchor steel/mm	Axle pressure/kN	Test axial compression ratio[a]	Note[b]
XJ	–	–	–	–	946.3	0.11	Contrast specimen
EHSW1	4	930	520.8	200	946.3	0.11	EHSW reference specimen
EHSW2	3	744	312.5	200	946.3	0.11	Pretension unchanged, the number of prestressed tendon changes
EHSW3	2	1116	312.5	200	946.3	0.11	
EHSW4	4	744	416.6	200	946.3	0.11	Prestressed tendon number unchanged, prestressed change
EHSW5	4	558	312.5	200	946.3	0.11	
EHSW6	4	930	520.8	250	946.3	0.11	Variation of unbonded length of grouted anchor bar
EHSW7	4	930	520.8	150	946.3	0.11	
EHSW8	4	930	520.8	200	473.15	0.055	Axial compression ratio change

Annotation [a] The axial compression ratio of the test is calculated according to the test axial pressure and the standard value of concrete axial compressive strength 23.4 MPa specified in the specification
[b] The comparison Specimens with different parameters can be combined again. Except for the change of axial compression ratio, there are only 2 comparison Specimens, and there are 3 comparison Specimens for the change of other parameters

blocks in each group, and the axial compression test is carried out before the specimen is loaded to obtain the measured value of the compressive strength of the concrete cube. According to the' Standard for evaluation of concrete compressive strength' (GB/T 50107-2010) [4], the statistical average of the cubic compressive strength of concrete materials is obtained. The compressive strength test of concrete test block was carried out on the pressure testing machine in the structural laboratory of Southeast University. Table 3.2 gives the measured values and related derived values of the performance indexes of concrete materials. Among them, the axial compressive strength and axial tensile strength and elastic modulus were calculated

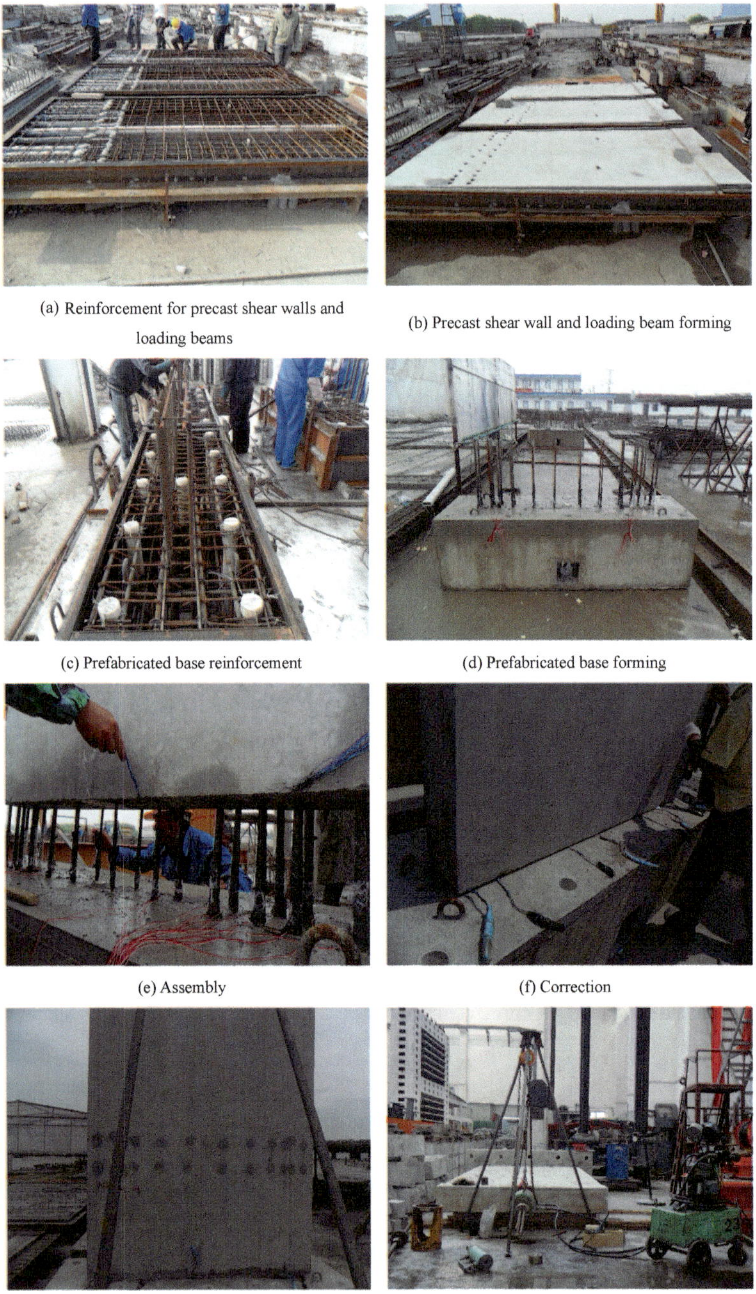

(a) Reinforcement for precast shear walls and loading beams

(b) Precast shear wall and loading beam forming

(c) Prefabricated base reinforcement

(d) Prefabricated base forming

(e) Assembly

(f) Correction

(g) Maintenance after grouting

(h) Steel strand tension

Fig. 3.6 EHSW specimen making process photos

Table 3.2 Concrete material performance index value of specimen

Concrete grades	Cube compressive strength 1/MPa	Axial compressive strength 2/MPa	Axial tensile strength 3/MPa	Elastic modulus 3/MPa
C35	35.4	23.7	2.5	31,444

Note [1] The statistical average of the measured values, according to the' concrete strength test evaluation standard' GB/T 50107-2010 4.3.1
[2] As the representative value of the compressive strength of concrete, it is used for the conversion of axial compression ratio and the determination of material parameters of subsequent finite element specimens
[3] For the determination of material parameters of subsequent finite element specimens

Table 3.3 Mechanical performance index value of steel bar material of specimen

Specifications for materials	Diameter/mm	Yield strength/MPa	Ultimate strength/ MPa	Elongation/%
HRB400	16	438	620	19
HRB400	10	430	557	14.3
HRB400	8	451	618	10.8
1860 grade steel strand	15.24	1650[1]	2050	5.6

Note [1] Since the steel strand material test curve has no obvious yield platform, the corresponding stress when the plastic strain is 0.2% is taken as the nominal yield stress

with reference to the "Code for design of concrete structures" (GB 50010-2010) [1] Article 4.1.3, Article 4.1.5 provisions described.

Reinforcement (ordinary steel, prestressed steel strand) according to the specifications of each reserved three samples to measure the yield strength of reinforced materials, ultimate strength and elongation. The tensile test of steel bar was carried out on the electro-hydraulic servo fatigue testing machine in the structural laboratory of Southeast University. The statistical average value of the mechanical performance index of steel bar is shown in Table 3.3.

3.2.4 Specimen Loading and Measurement

1) Test loading device

This test adopts the quasi-static loading scheme, that is, under constant axial compression, horizontal low cyclic loading is applied at the center of the loading beam section at the top of the specimen. The test loading device is shown in Fig. 3.7.

As shown in Fig. 3.7, the test loading device is basically divided into three parts: vertical loading system, horizontal loading system and anchorage system.

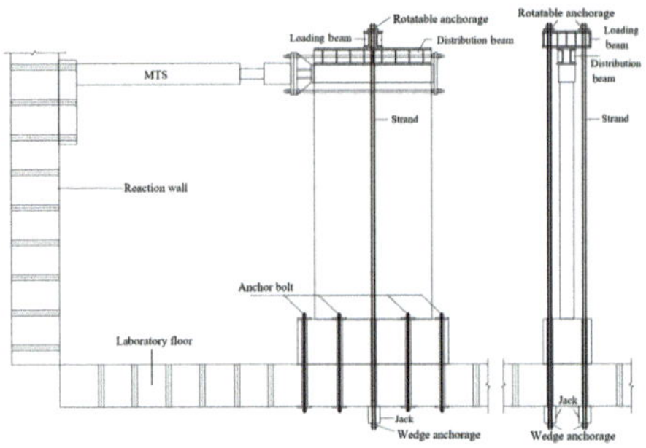

(a) Schematic plan

(b) On-site photographs

Fig. 3.7 Test loading device

In order to apply the predetermined axial pressure, the vertical loading system adopts the external prestressing loading scheme. The loading device is composed of external prestressing equipment, loading beam and distribution beam. Four 1860 Φ15.24 prestressed steel strands were set on both sides of the middle of the specimen. Two 100 t tension jacks were placed on the loading beam for convenient installation. The applied vertical force was transmitted to the distribution beam with relatively large stiffness through the loading beam, and then evenly transmitted to the top surface of the specimen. In addition, in order to adapt to the horizontal deformation of the specimen during the loading process and avoid the torsion or shear of the steel strand, a rotatable anchor is used at the top so that the steel strand can achieve a certain degree of free rotation with the horizontal deformation of the specimen.

The horizontal loading system adopts the 1000 kN electro-hydraulic servo actuator made by MTS company. The actuator tail is anchored on the reaction wall through the steel plate welding base and bolt. The hand-pulled gourd is pulled obliquely on the reaction wall near the expansion part of the oil cylinder to adjust the installation position of the actuator and prevent the sudden accident of the actuator falling; the actuator loading head is connected with the specimen loading beam by 4 Φ32 finishing screw thread steels and splints. At the same time, considering that the vertical loading steel strand may collide or rub with the horizontal finishing rolled rebar, a round steel tube is sleeved in the middle of the finishing rolled rebar to reduce the mutual interference between the two.

The anchorage system includes two parts, one is the connection between the specimen and the laboratory ground. In order to simulate the fixed boundary conditions at the bottom of the Specimen, eight Φ32 finishing rolled rebars were used to pass through the reserved holes in the base of the Specimen and the corresponding holes on the laboratory ground, and the finishing rolled rebars were tensioned (the tension force was about 20 t). The Specimen was pressed with the laboratory ground to avoid the rigid body rotation and horizontal slip of the Specimen. The other part is the out-of-plane restraint device of the specimen. In the previous similar tests of the research group, it was inevitable that the wall torsion occurred in the assembled concrete shear wall specimens in the later stage of loading, which caused the premature failure of the assembled shear wall specimens to a certain extent and affected the test results. In order to overcome this problem and ensure the authenticity of the test results, at the same time, in order to ensure the safety of the test process and prevent the lateral overturning of the specimen, two triangular steel trusses were installed on both sides of the specimen. The steel trusses were placed on the wall of the specimen through the roller to limit its in-plane rotation and lateral deformation.

2) Test loading system

According to the' Specification for seismic test of buildings' (JGJ/T 101–2015) [5], the loading system of load–displacement double control is adopted. In the load control stage, the first stage of 50 kN (individual loading level is subdivided into 25 kN), each stage is cycled once; displacement control stage, each level cycle 3 times, at the same time, in order to facilitate the comparison of each specimen, all specimens of the loading displacement of 23 mm as a level; conversion from load control to displacement control is based on the first yield of vertical steel bars. The specific loading system is shown in Fig. 3.8, where the actuator extrapolation is positive and the inner pull is negative.

The specific loading program of the test is:

(1) Test axial compression. 30% of the vertical force is applied at the top of the specimen, then unloaded and repeated 2–3 times. The symmetry of the specimen loading is observed by the steel strain gauge, and the vertical gap between the specimen and the loading device is eliminated.
(2) Apply constant axial pressure. Apply axial pressure to the predetermined axial pressure at one time and remain unchanged.

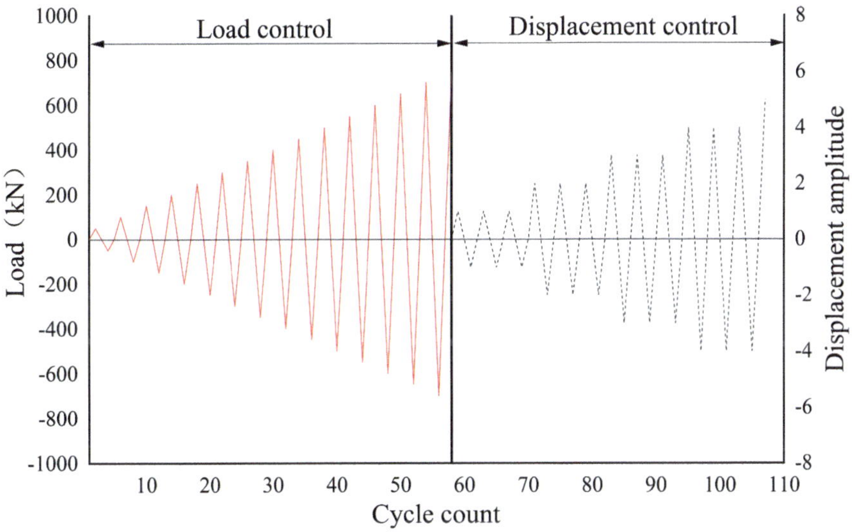

Fig. 3.8 Test loading system

(3) Apply horizontal repeated load. A 10 kN horizontal load is applied by the actuator and a cycle is performed. The symmetry of the loading response of the specimen is judged by the hysteresis curve collected by the actuator and the gap between the specimen and the loading device is eliminated.
(4) Officially start the test. According to the loading system, horizontal repeated loads are applied until the specimen is damaged or is not suitable for further loading.

3) Test measurement scheme

(1) Axial force at the top of the specimen. The actual vertical axial force can be calculated according to the reading of the precision pressure dial and the calibration curve by counting the oil pressure gauge of the oil pressure jack.
(2) Load–displacement hysteresis curve of loading point at the top of specimen. Measured directly by MTS actuators.
(3) Steel strain. Including the shear wall edge member vertical reinforcement and stirrup strain and reserved vertical reinforcement strain. The corresponding strain is measured by pasting a resistance strain gauge on the steel bar.
(4) Specimen prestressed reinforcement internal force. The internal force of unbonded prestressed tendon is measured by pre-set anchor cable pressure sensor.
(5) Crack width. Using special crack width reading microscope measurement.

4) Axial compression ratio conversion

According to the axial pressure applied in the test and the measured values of the mechanical properties of the material, considering the load and material partial coefficients, the test axial compression ratio in Table 3.1 is converted into the design axial compression ratio. The conversion process is as follows:

Considering the load partial factor, the conversion relationship between the test axial force and the gravity load representative value in the calculation of the axial compression ratio of the shear wall is shown in Eq. 3.2, where 1.2 is the partial factor of the gravity load representative value.

$$N = 1.2N_t \tag{3.2}$$

Considering the partial coefficient of material, the conversion relationship between the measured value and the design value of concrete strength is shown in Formula 3.3, where 1.4 is the partial coefficient of concrete material, and σ_{fc} is the coefficient of variation of concrete strength, taken as 0.13 (according to the superseded' Code for design of concrete structures' (GB 50010–2002)).

$$f_c = \frac{1 - 1.645\delta_{fc}}{1.4} f_{c,t} \tag{3.3}$$

The conversion formula between test axial compression ratio and design axial compression ratio is shown in Eq. 3.4.

$$\mu = \frac{N}{f_c bh} = \frac{1.2N_t}{\frac{1-1.645\delta_{fc}}{1.4} f_{c,t} bh} = 2.14\mu_t \tag{3.4}$$

Therefore, the axial compression ratio of 0.110 and 0.055 involved in this test should be converted to the design axial compression ratio of 0.24 and 0.12 respectively.

3.3 Test Process and Test Phenomenon

3.3.1 Failure Process of Specimen

The failure process of each specimen is essentially the same, comprising several main stages including elasticity, cracking, yielding, and failure. The specific failure process of each specimen is detailed below.

1) Specimen XJ

The main phenomena observed during the XJ loading of the specimen are shown in Table 3.4 and the local damage is shown in Fig. 3.9.

2) Specimen EHSW1

The main phenomena observed during the EHSW1 loading process of the specimen are presented in Table 3.5, and the corresponding local damage is depicted in Fig. 3.10.

3) Specimen EHSW2

The main phenomena observed during the EHSW2 loading process of the specimen are summarized in Table 3.6, and the corresponding local damage is depicted in Fig. 3.11.

4) Specimen EHSW3

The main phenomena observed during the EHSW3 loading process of the specimen are shown in Table 3.7, and the local damage is depicted in Fig. 3.12.

5) Specimen EHSW4

The main phenomena observed in the EHSW4 loading process of the specimen are presented in Table 3.8, and the corresponding local damage is illustrated in Fig. 3.13.

Table 3.4 Specimen XJ loading process

Loading process		Main phenomena
Load control	Elastic stage (0 ~ ± 225 kN)	The load and displacement change linearly; the loading and unloading curves essentially coincide; and there is virtually no residual deformation after unloading
	Cracking stage(± 225 kN)	Horizontal cracks appear on the top surface of the specimen base, or root of the wall, with a maximum crack width of approximately 0.01 mm
	Yielding phase(± 500 kN)	The vertical steel bar at the edge of the tensile side of the wall yields, and the average value of forward and reverse horizontal displacement of the specimen is 23 mm[a]. New cracks continuously form along the height of the wall edges, with the highest crack located approximately 2.0 m from the top of the base. The crack is shorter in the horizontal section and develops faster obliquely downward to the neutral axis. From the bottom of the specimen, positive and negative diagonal cracks have crossed the neutral axis after intersection, forming cross diagonal cracks in a height range of about 50 cm from the base In this stage, the maximum crack width increases continuously and stably, and the maximum crack width is about 0.44 mm
Displacement control	Displacement loading stage	After entering the displacement loading stage, several new cracks appear along the height direction of the wall, with only one new crack developing in both the positive and negative directions at the top of the wall. The cracks primarily continue to extend on existing cracks and intersect over the neutral axis. Positive and negative cross diagonal cracks have formed in the height range of approximately 1.5 m from the top surface of the base When loaded to ± 46 mm stage, vertical cracks began to appear in the concrete protective layer. Loaded to ± 69 mm stage, the protective layer and the core concrete completely separated, with local spalling of the protective layer concrete and corner crushing of the core concrete When loaded to ± 46 mm stage, the maximum crack width increases sharply from 0.44 mm to 2.2 mm

(continued)

Table 3.4 (continued)

Loading process		Main phenomena
	Destruction phase (± 92 mm)	During the second loading cycle of this stage, the outermost tensile steel bar of the specimen broke individually, the steel bar on the compression side was obviously convex and buckled, the protective layer concrete peeled off in a large range, the internal vertical steel bars and stirrups were exposed, and the core concrete was locally compressed Damage to the concrete is concentrated at both ends of the wall limb, with a damage range of approximately 20–30 cm from the top of the base in the height direction

Note [a] This value serves as the basis for the unit loading displacement of the testing system in this experiment

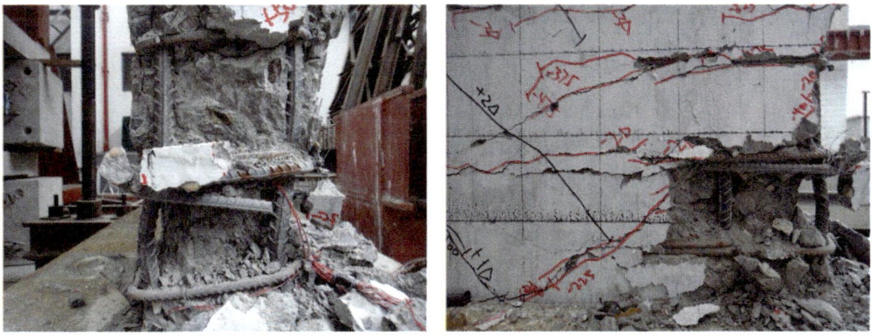

(a) wall side (b) wall front

Fig. 3.9 Local failure of specimen XJ

6) Specimen EHSW5

The main phenomena observed in the EHSW5 loading process of the specimen are shown in Table 3.9, and the local damage is shown in Fig. 3.14.

7) Specimen EHSW6

The main phenomena observed in the EHSW6 loading process of the specimen are shown in Table 3.10, and the local damage is shown in Fig. 3.15.

8) Specimen EHSW7

The main phenomena observed in the loading process of the specimen EHSW7 are shown in Table 3.11, and the local damage is shown in Fig. 3.16.

Table 3.5 Specimen EHSW1 loading process

Loading process		Main phenomena
Load control	Elastic stage (0 ~ ± 300 kN)	The load and displacement change linearly; the loading and unloading curves essentially coincide, and there is minimal residual deformation after unloading
	Cracking stage (± 300 kN)	Horizontal cracks appear on the top surface of the specimen base and the upper layer of the slurry layer, with a maximum crack width of approximately 0.01 mm
	Yielding phase (± 550 kN)	The vertical reinforcement at the most edge of the tensile side of the wall has yielded; the edge of the wall continues to produce new cracks along the height direction. The highest crack is approximately 1.5 m from the top surface of the base; the higher the crack, the shorter the horizontal section, and the faster it develops downward to the neutral axis. Compared to specimen XJ, the crack tilts earlier and the tilt angle is larger. In the height range of about 70 cm from the base, the positive and reverse diagonal cracks at the bottom of the specimen have crossed the neutral axis and intersected to form cross diagonal cracks. Different from specimen XJ, the cracks at the bottom of the specimen are less developed, and there are no cracks in the range of about 150 mm from the top of the base at both ends of the wall During this stage, the cracks continued to develop steadily, with a maximum crack width of approximately 0.22 mm. The seam also exhibited opening and closing, with a maximum opening value of about 1.2 mm
Displacement control	Displacement loading stage	Upon entering the displacement loading stage, the wall continues to develop new cracks along the height direction, with the highest crack occurring approximately 2.3 m from the top surface of the base. Building on existing cracks, oblique downward extensions continue, generating positive and negative cracks that cross the neutral axis intersection. From the top base surface, oblique cracks of approximately 1.6 m height range have been formed, spanning both positively and negatively When loaded to ± 46 mm stage, vertical cracks started appearing in the concrete protective layer. When loaded to the ± 69 mm stage, the protective layer separated from the core concrete, and the concrete of the local protective layer peeled off, while the exposed core concrete remained intact Loading to ± 46 mm stage, the maximum crack width increased from 1.3 to 3.2 mm; abutment opening is stable growth. Loading to ± 92 mm stage, the abutment maximum opening value is about 20 mm

(continued)

Table 3.5 (continued)

Loading process		Main phenomena
	Destruction phase (± 92 mm)	Upon loading the third cycle of this stage, the outermost tensile steel bar of the specimen was individually broken, the outer corrugated pipe was torn, the compressive side steel bar was visibly convex and buckled, the protective layer concrete was widely peeled off, the internal vertical steel bar and stirrups were exposed, and the core concrete was partially crushed The damage to the concrete is concentrated at both ends of the wall limb, with a damage range of approximately 10–20 cm from the top of the base in the height direction. The damage is less severe than that observed in specimen XJ

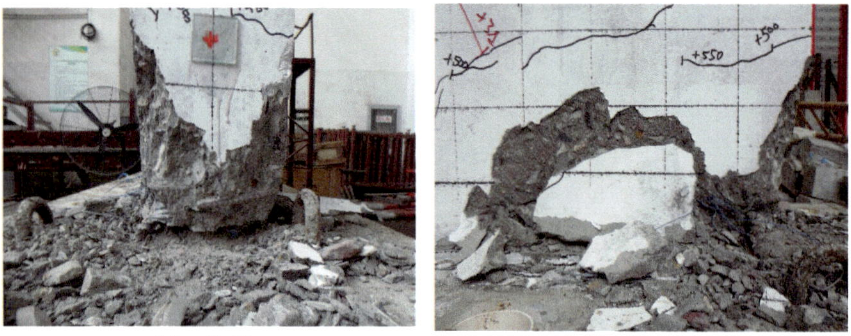

(a) wall side (b) wall front

Fig. 3.10 Local failure of EHSW1 specimen

9) Specimen EHSW8

The main phenomena observed in the EHSW8 loading process of the specimen are shown in Table 3.12, and the local damage is shown in Fig. 3.17.

3.3.2 Failure Form of Specimen

Similar to a cantilever force member, a shear wall can fail in various modes including bending, bending-shear, and sheer failure. The shear span ratio is a critical parameter determining the failure mode of a shear wall. Generally, when the shear span ratio exceeds 2, the shear wall section becomes predominantly flexural and prone to flexural failure. The cracks in the shear wall are primarily horizontal, with a small number of oblique cracks appearing near the mid-wall. When the shear span ratio is between 1 and 2, the section of the shear wall is controlled by both the bending moment and the shear force. The cracks in the shear wall remain horizontal bending cracks; however, under the additional action of the shear force, the cracks incline downward towards the neutral axis of the wall, forming bending shear diagonal cracks. Under reciprocating loads, the diagonal cracks intersect at the neutral axis, creating cross diagonal cracks. When the shear span ratio is less than 1, the shear wall section mainly suffers shear, prone to shear failure. The shear wall crack manifests as a wide shear oblique crack. Typical shear failure modes include diagonal tension failure, diagonal compression failure, and shear slip failure.

Figure 3.18 presents the final crack distribution for each specimen after testing. As can be seen from the figure, like specimen XJ, all EHSW specimens experienced flexural-shear failure. The shear wall body featured a curved horizontal crack and a shear diagonal crack; these diagonal cracks were staggered at the neutral axis to form cross cracks. Meanwhile, from the test process description, it can be inferred that each specimen failed in an ideal reinforced manner, that is, the vertical reinforcement of the shear wall was broken, which led to specimen failure. Although the core concrete

Table 3.6 EHSW2 loading process of specimen

Loading process		Main phenomena
Load control	Elastic stage (0 ~ ± 250 kN)	The load and displacement change linearly, and the loading and unloading curves essentially coincide, with little to no residual deformation after unloading
	Cracking stage (± 250 kN)	Horizontal cracks appear on the top surface of the specimen base and upper layer of the slurry layer, with a maximum crack width of about 0.01 mm
	Yielding phase (± 525 kN)	The vertical reinforcement at the top edge of the tensile side of the wall yielded; the edge of the wall, along the height direction, generated new cracks continuously. The highest crack initiated from the top base surface, approximately 1.7 m, and the cracks developed upward, with the shorter horizontal section, and faster near the neutral axis obliquely downward. Compared with specimen XJ, the cracks tilted earlier and at a greater angle; within the height range of about 70 cm from the base, the positive and reverse diagonal cracks at the bottom of the specimen had crossed the neutral axis and intersected to form cross diagonal cracks. Similarly, obviously differing from specimen XJ, the cracks at the bottom of the specimen were less developed, and there were no cracks in the range of about 200 mm from the top base at both ends of the wall, with the existing cracks close to the middle of the wall, which was the extension of the oblique cracks in the upper part of the wall At this stage, the crack develops continuously and stably, with a maximum width of approximately 0.32 mm. Open and closed joints also appear, with a maximum opening value of about 1.2 mm
Displacement control	Displacement loading stage	Upon entering the displacement loading stage, the wall continues to develop new cracks along the height direction; the highest crack is approximately 2.6 m from the top surface of the base. On the basis of existing cracks, they continue to extend obliquely downward. Positive and reverse cracks intersect over the neutral axis, and within a height range of about 1.4 m from the top surface of the base, positive and reverse cross-oblique cracks have formed When loaded to ± 46 mm stage, vertical cracks began to appear in the concrete protective layer. When loaded to the ± 69 mm stage, the protective layer separated from the core concrete, and the concrete of the local protective layer peeled off, while the exposed core concrete remained basically intact During the loading stage of ± 46 mm, the maximum crack width increases from 1.8 to 3.5 mm. The abutment opening exhibits stable growth; by the loading stage of ± 92 mm, the abutment maximum opening value is approximately 18 mm

(continued)

Table 3.6 (continued)

Loading process		Main phenomena
	Destruction phase (± 92 mm)	Loaded to the first cycle of this stage, the outermost tensile steel bar of the specimen was broken individually, exposing the bellows which were torn, the compressive side steel bar was obviously convex and buckled, the protective layer concrete was peeled off in a large range, exposing the internal vertical steel bar and stirrups, and the core concrete was partially crushed Damage to the concrete is concentrated at both ends of the wall limb, with a damage range of approximately 10–20 cm from the top of the base in the height direction. The damage is less severe than that observed in specimen XJ It was observed that torsion of the specimens during loading was very pronounced, which may lead to under-exploitation of specimen properties

(a) wall side (b) wall front

Fig. 3.11 Local failure of EHSW2 specimen

was crushed to some degree, it still exhibited sufficient deformation capacity and bearing strength.

The specific analysis of the relevant test phenomena is as follows:

(1) For the EHSW specimen, an important phenomenon different from specimen XJ is the absence of cracks at both ends of the wall within a certain height range from the top of the base (the distance from the top of the base is between 150 and 300 mm). Since there is no stress transfer between the steel bar and the concrete in the unbonded section, the tensile stress transmitted by the steel bar does not manifest in the concrete at this part, thereby preventing the occurrence of concrete cracks [6]. However, by comparing EHSW specimens with different unbonded lengths (EHSW7 (150 mm), EHSW1 (200 mm), EHSW6 (250 mm)), no obvious regularity between the height range of crack-free and unbonded length is found (maximum crack-free heights of the three specimens are 200 mm, 150 mm, and 220 mm respectively). This may be mainly due to the discreteness of concrete material properties, especially for tensile properties of concrete, which is particularly prominent. In addition, specimen manufacturing error and possible slurry leakage in the plastic film wrapping of the unbonded section can also change the no-crack range [7].
(2) For the EHSW specimen, another important phenomenon that differs from the XJ specimen is that, except for EHSW8, the horizontal crack section of the other EHSW specimens is shorter, and the inclined crack angle is larger. The analysis shows that the existence of pretension in unbonded prestressed tendons essentially increases the vertical compressive stress in the concrete of the specimen. The increase of vertical compressive stress will lead to a change in the principal stress direction of the concrete, and for the EHSW specimen, the diagonal cracks will tilt downward faster. For the EHSW8 specimen, although there is still the influence of preloading stress, the influence of vertical force is weakened due to its low axial pressure. Therefore, the crack shape of this specimen is similar to that of the XJ specimen [6].

Table 3.7 Specimen EHSW3 loading process

Load process		Main phenomena
Load control	Elastic stage (0 ~ ± 250 kN)	The load and displacement change linearly; the loading and unloading curves are essentially coincident; and there is virtually no residual deformation after unloading
	Cracking stage (± 250 kN)	Horizontal cracks appear on the top surface of the specimen base and the upper layer of the slurry layer, with a maximum crack width of about 0.01 mm
	Yielding phase (± 500 kN)	The vertical reinforcement at the top edge of the tensile side of the wall has yielded; the edge of the wall continues to produce new cracks along the height direction. The highest crack is about 1.6 m from the top surface of the base, and the higher the crack, the shorter the horizontal section, and the faster it develops downward to the neutral axis. Compared with specimen XJ, the crack tilts earlier and the tilt angle is larger. In the height range of about 70 cm from the base, the positive and reverse diagonal cracks at the bottom of the specimen have crossed the neutral axis and intersected to form cross diagonal cracks. Similarly, obviously different from specimen XJ, the cracks at the bottom of the specimen are less developed, and there are no cracks in the range of about 220 mm from the top of the base at both ends of the wall, and the existing cracks are close to the middle of the wall limb, which is the extension of the oblique cracks in the upper part of the wall In this stage, the cracks continue to develop steadily, and the maximum crack width is about 0.26 mm. Joints also appear open and closed phenomenon, the maximum opening value of about 1.6 mm
Displacement control	Displacement loading stage	Entering the displacement loading stage, the wall continues to develop new cracks along the height direction; the highest crack is about 2.5 m from the top surface of the base. On the basis of existing cracks, it continues to extend obliquely downward. The positive and reverse cracks intersect over the neutral axis, and the positive and reverse cross-oblique cracks within a height range of about 1.5 m from the top surface of the base have formed When loaded to ± 46 mm stage, vertical cracks began to appear in the concrete protective layer. When loaded to the ± 69 mm stage, the protective layer separated from the core concrete, and the concrete of the local protective layer peeled off, while the exposed core concrete remained basically intact Loading to ± 69 mm stage, the maximum crack width increased from 0.66 to 2.0 mm; abutment opening is stable growth. Loading to ± 92 mm stage, the abutment maximum opening value is about 23 mm

(continued)

Table 3.7 (continued)

Load process		Main phenomena
	Destruction phase (± 92 mm)	Upon loading to the second cycle of this stage, the outermost tensile steel bar of the specimen was individually broken, exposing the bellows which was torn, the compressive side steel bar was noticeably convex and buckled, the protective layer concrete was peeled off in a large range, the internal vertical steel bar and stirrups were exposed, and the core concrete was partially crushed The damage to the concrete is concentrated at both ends of the wall, with a damage range of approximately 10–15 cm from the top surface of the base in the height direction. The damage is less severe than that observed in Specimen XJ

(a) wall side (b) wall front

Fig. 3.12 Local failure of specimen EHSW3

(3) For the EHSW specimen, an important phenomenon different from the XJ specimen is that damage is more concentrated near the abutted seam and lighter overall. Analysis shows that mid-to-late stage loading inhibits horizontal joint opening, limiting plastic hinge development. Light damage results from tensile crack inhibition at the specimen end, ensuring concrete compression integrity and improving bearing and deformation capacities, thus reducing damage [8]. Moreover, wall end buckled closed stirrups further encrypt the end, effectively improving restraint and deformation, and reducing damage.

(4) For the EHSW specimens with different unbonded lengths, no anchorage slip failure occurs, and the tensile steel bars are broken. Therefore, it can be considered that for Φ16 steel bar (HRB400), the slurry anchor connection is adopted, and the anchorage length of 350 mm can ensure the full play of the tensile force of the steel bar. The length is about 22d, which is about 60% of its seismic anchorage length of 37d (according to the first and second seismic, C35 calculation), which is basically consistent with the previous test results of the research group.

(5) It should be noted that for EHSW specimens, although the anti-torsion measures are considered in the test device, wall torsion is still observed, especially in the later stage of loading. Analysis indicates that in the later stage of loading, the slurry layer is crushed under the combined action of compression and shear, but the degree of crushing is not uniform throughout the section, resulting in the bottom of the shear wall, particularly the end of the wall limb, acting as an uneven rotating fulcrum. This inevitably leads to torsion occurrence. Such phenomenon will cause EHSW specimens to underperform, limiting their performance potential and rendering reliable speculation inconclusive. Based on test results, EHSW specimens should exhibit better performance.

Table 3.8 Loading process of specimen EHSW4

Loading process		Main phenomena
Load control	Elastic stage (0 ~ ± 275 kN)	The load and displacement change linearly; the loading and unloading curves are essentially coincident; and there is virtually no residual deformation after unloading
	Cracking stage (± 275 kN)	Horizontal cracks appear on the top surface of the specimen base and the upper layer of the slurry layer, with a maximum crack width of about 0.01 mm
	Yielding phase (± 550 kN)	The vertical reinforcement at the top edge of the tensile side of the wall yielded; the wall edge along the height direction generated new cracks continuously, with the highest crack from the top base surface at approximately 1.7 m, and shorter, faster cracks developing obliquely downward toward the neutral axis, compared with specimen XJ, with earlier tilt of cracks and greater tilt angle; within the height range of about 70 cm from the base, the positive and reverse diagonal cracks at the bottom of the specimen crossed the neutral axis and intersected to form cross diagonal cracks. Similarly, unlike specimen XJ, the bottom cracks were less developed. There were no cracks in the range of about 250 mm from the top base at both ends of the wall, with existing cracks close to the middle of the wall limb, extending inclined cracks on the upper wall At this stage, the crack develops continuously and stably, with a maximum crack width of about 0.3 mm. The seam also exhibits opening and closing, with a maximum opening value of about 1.5 mm
Displacement control	Displacement loading stage	After entering the displacement loading stage, the wall continues to develop new cracks along the height direction. The highest crack is approximately 2.3 m from the top surface of the base and continues to extend obliquely downward. The positive and reverse cracks intersect over the neutral axis, and within a height range of about 1.5 m from the top surface of the base, the positive and reverse cross-oblique cracks have been formed When loaded to ± 46 mm stage, vertical cracks began to appear in the concrete protective layer. When loaded to the ± 69 mm stage, the protective layer separated from the core concrete, and the concrete of the local protective layer peeled off, while the exposed core concrete remained basically intact During the loading stage of ± 69 mm, the maximum crack width increases from 0.58 mm to 3.0 mm. The abutment opening exhibits stable growth; by the loading stage of ± 92 mm, the abutment maximum opening value is approximately 20 mm

(continued)

Table 3.8 (continued)

Loading process		Main phenomena
	Destruction phase (± 92 mm)	Upon loading to the third cycle of this stage, the outermost tensile steel bar of the specimen was individually broken, exposing the bellows which was torn, the compressive side steel bar was visibly convex and buckled, the protective layer concrete was extensively peeled off, revealing the internal vertical steel bars and stirrups, and partially crushing the core concrete The damage to the concrete is concentrated at both ends of the wall, with the damage range being approximately 10 ~ 30 cm from the top surface of the base in the height direction. The damage is less severe than that observed in specimen XJ

(a) wall side (b) wall front

Fig. 3.13 Local failure of EHSW4 specimen

3.4 Hysteresis Curve Analysis

The load–displacement curve for the entire cyclic loading process is the hysteresis curve, reflecting the hysteresis characteristics of the reinforced concrete structure or component due to the material's nonlinear properties. Unloading to zero load does not restore the displacement, indicating residual deformation. It comprehensively reflects the structure's or component's bearing capacity, stiffness, deformation capacity, and energy dissipation capacity under cyclic loading. Hysteresis curves can be divided into four basic forms based on their shape: spindle, bow, inverse S, and Z (see Fig. 3.19) [9]. The shuttle-shaped hysteresis curve is full in shape, without pinching effect or shear slip, reflecting strong elastic–plastic performance of the structure or component. The structure or component with this type of curve has good seismic performance and energy dissipation capacity. It is common in bending, eccentric compression, and bending shear members without shear failure. The arch-shaped hysteresis curve has a relatively full shape, with pinching and a certain shear slip, reflecting strong elastic–plastic performance. The structure or component with this type of curve has better seismic performance and energy dissipation capacity. It is common in flexural-shear members and compressive-flexural-shear members with large shear span ratio, small shear force and certain stirrups. In general reinforced concrete structures, the hysteresis curve belongs to this type; the inverse S-shaped hysteresis curve is not full enough, and the pinch effect is obvious, reflecting more shear slip effects. The seismic performance and energy dissipation capacity of the structure or component with this type of curve are poor, common in general frames, beam-column joints and shear walls. The Z-shaped hysteresis curve reflects a large number of shear slip effects. The curve has slip properties, generally shear failure or anchorage failure mode. It is common in components with small shear span and fully developed oblique cracks and components with large slip of anchorage steel bars.

The load–displacement hysteresis curves for each specimen are presented in Fig. 3.20. The diagrams reveal that the hysteresis curve characteristics for the

Table 3.9 EHSW5 loading process of specimen

Loading process		Main phenomena
Load control	Elastic stage(0 ~ ± 250 kN)	The load and displacement change linearly; the loading and unloading curves essentially coincide; and there is essentially no residual deformation after unloading
	Cracking stage (± 250 kN)	Horizontal cracks appear on the top surface of the specimen base and the upper layer of the slurry layer, with a maximum crack width of about 0.01 mm
	Yielding phase (± 550 kN)	The vertical reinforcement at the top edge of the tensile side of the wall yielded; the wall edge along the height direction continued to generate new cracks, with the highest crack initiating from the top base surface at about 1.9 m, and the cracks propagating upward and shorter in the horizontal section, faster towards the neutral axis obliquely downward, compared with specimen XJ, exhibiting earlier crack tilt and greater tilt angle; the positive and reverse diagonal cracks at the bottom of the specimen crossed the neutral axis within the height range of about 90 cm from the base, forming cross diagonal cracks. Similarly, obviously different from specimen XJ, the cracks at the bottom of the specimen were less developed, with no cracks in the range of about 300 mm from the top base surface at both ends of the wall During this stage, the cracks continued to develop steadily, with a maximum width of approximately 0.44 mm. Joints also appear open and closed phenomenon, the maximum joint opening is about 1.9 mm
Displacement control	Displacement loading stage	After entering the displacement loading stage, the wall continues to develop new cracks along the height direction, with the highest crack located about 2.3 m from the top surface of the base. On the basis of the existing cracks, oblique downward extension continues, resulting in positive and negative cracks crossing the neutral axis intersection. From the top base surface, oblique cracks of approximately 1.6 m height have been formed in the positive and negative cross directions When loaded to ± 46 mm stage, vertical cracks began to appear in the concrete protective layer. When loaded to the ± 69 mm stage, the protective layer separated from the core concrete, and the concrete of the local protective layer peeled off; meanwhile, the exposed core concrete remained basically intact During the loading stage of ± 69 mm, the maximum crack width increases from 0.58 mm to 3.0 mm. The abutment opening exhibits stable growth; by the loading stage of ± 92 mm, the abutment maximum opening value reaches approximately 23 mm

(continued)

Table 3.9 (continued)

Loading process		Main phenomena
	Destruction phase (± 92 mm)	When loaded to the second cycle of this stage, the outermost tensile steel bar of the specimen was broken individually, the exposed bellows were torn, the compressive side steel bar was obviously convex and buckled, the protective layer concrete was peeled off in a large range, the internal vertical steel bar and stirrup were exposed, and the core concrete was partially crushed Concrete damage is concentrated at both ends of the wall, with damage extending approximately 10–30 cm from the top surface of the base in the height direction. The damage is less severe than that observed in specimen XJ

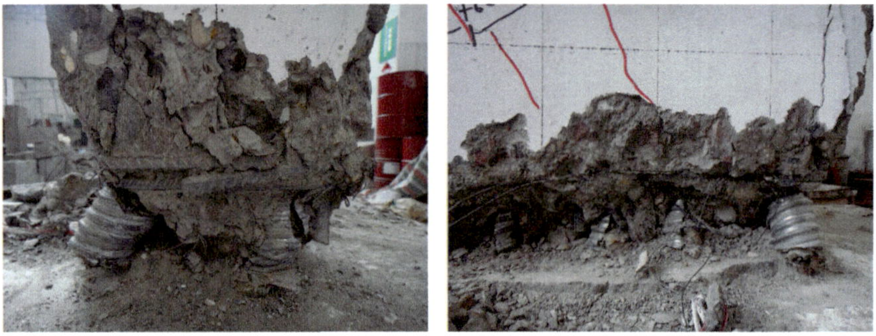

(a) wall side (b) wall front

Fig. 3.14 Local failure of specimen EHSW5

EHSW and XJ specimens are essentially the same. Prior to cracking, the loading and unloading curves are essentially coincident, the hysteresis loop area is small, the curve is essentially linear, and the specimen is in an elastic working state. After cracking, the stiffness of the specimen decreases, the curve inclines toward the displacement axis, and there is still residual deformation after unloading, indicating that the specimen has entered the elastic–plastic working state, and the hysteresis loop begins to bend and develop steadily. The residual deformation of the specimen increases slowly, the area of the hysteresis loop increases gradually, and the hysteresis curve assumes a sharp spindle shape. In the displacement loading stage, the shear displacement component increases, and the effect of shear on the pinching of the hysteresis curve enhances. Meanwhile, the bond slip develops rapidly after the steel bar yields, and the pinching effect of the hysteresis curve caused by slip increases, making the pinching effect of the hysteresis curve increasingly evident. The residual deformation of the specimen grows, and the shape of the hysteresis loop changes directly from the sharp spindle to the anti-S shape. With the increase of the loading displacement, the shape of the hysteresis curve remains as an inverse S shape, but the curve is full, reflecting the good elastic–plastic working performance of the specimen. For the cyclic loading conditions experienced by the final failure of each specimen, the specimens XJ, EHSW3, EHSW5, EHSW6, and EHSW7 all completed the second cycle of the fourth-level displacement loading level. The specimens EHSW1, EHSW4, and EHSW8 all completed the third cycle of the fourth-level displacement loading level, and the specimen EHSW2 only completed the first cycle of the fourth-level displacement loading level.

The hysteretic curves for different unbonded prestressed tendon pretension (tension control stress), different unbonded prestressed tendon area (number), different unbonded length of slurry anchor steel, and different axial compression ratio are depicted in Fig. 3.21.

Table 3.10 EHSW6 loading process

Loading process		Main phenomena
Load control	Elastic stage (0 ~ ± 300 kN)	The load and displacement change linearly; the loading and unloading curves are essentially coincident; and there is virtually no residual deformation after unloading
	Cracking stage (± 300 kN)	Horizontal cracks appear on the top surface of the specimen base and the upper layer of the slurry layer, with a maximum crack width of approximately 0.01 mm
	Yielding phase (± 575 kN)	The vertical reinforcement at the top edge of the tensile side of the wall yielded; the edge of the wall continued to produce new cracks along the height direction. The highest crack was about 1.9 m from the top surface of the base, and the higher the crack, the shorter the horizontal section, and the faster it developed downward to the neutral axis. Compared with specimen XJ, the crack tilted earlier and at a larger angle. In the height range of about 70 cm from the base, the positive and reverse diagonal cracks at the bottom of the specimen had crossed the neutral axis and intersected to form cross diagonal cracks. Similarly, unlike specimen XJ, the cracks at the bottom of the specimen were less developed. There were no cracks in the range of about 220 mm from the top surface of the base at both ends of the wall, and the existing cracks were close to the middle of the wall limb, which was the extension of the inclined cracks in the upper part of the wall At this stage, the cracks continue to develop stably, reaching a maximum width of approximately 0.46 mm. Joints also appear open and closed phenomenon, the maximum joint opening is about 2.0 mm
Displacement control	Displacement loading stage	After entering the displacement loading stage, the wall continues to develop new cracks along the height direction, with the highest crack located approximately 2.4 m from the top surface of the base. On this basis, existing cracks continue to extend obliquely downward, forming both positive and negative cracks that cross the neutral axis intersection. From the top of the base surface, positive and negative cross oblique cracks have been formed in a height range of about 1.6 m When loaded to ± 46 mm stage, vertical cracks began to appear in the concrete protective layer. When loaded to the ± 69 mm stage, the protective layer separated from the core concrete, and the concrete of the local protective layer peeled off, while the exposed core concrete remained basically intact Loading to ± 69 mm stage, the maximum crack width increased from 0.7 mm to 3.2 mm; abutment opening exhibited stable growth. Loading to ± 92 mm stage, the abutment maximum opening value reached approximately 21 mm

(continued)

Table 3.10 (continued)

Loading process		Main phenomena
	Destruction phase (± 92 mm)	When loaded to the second cycle of this stage, the outermost tensile steel bar of the specimen was broken individually, exposing the bellows, which were torn; the compressive side steel bar was obviously convex and buckled; the protective layer concrete was peeled off in a large area, exposing the internal vertical steel bar and stirrups; and the core concrete was partially crushed The damage of the concrete is concentrated at both ends of the wall limb, and the damage range is about 10–20 cm from the top of the base in the height direction, and the damage degree is slighter than that of the specimen XJ

(a) wall side

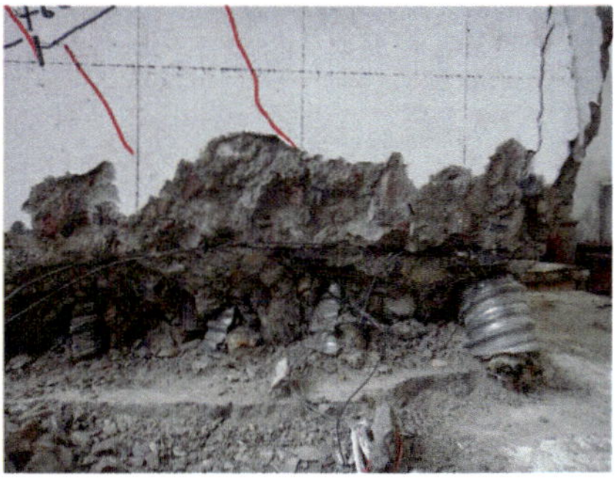

(b) wall front

Fig. 3.15 Local failure of EHSW6 specimen

3.5 Skeleton Curve Analysis

The outer envelope of the hysteresis curve is the skeleton curve, which is the trajectory of the maximum peak value of the horizontal force reached by each cyclic loading. It reflects the different stages and characteristics of the stress and deformation of the component (strength, stiffness, ductility and energy dissipation, etc.), and is also an important basis for determining the characteristic points in the restoring force model.

The skeleton curves of each specimen are shown in Fig. 3.22. From the diagram, it can be found that the skeleton curve of each specimen is basically the same, showing similar development rules. Under low cyclic loading, each specimen experienced several stages of elasticity, cracking, yield, limit, and failure. After yield, the skeleton curve continues to rise slowly, indicating that the bearing capacity of the specimen still has a certain degree of improvement. After approaching the peak load, the curve is basically flat, and there is no obvious sudden drop in bearing capacity. It shows that under the condition of large deformation in the later stage of loading, the specimen can still maintain sufficient bearing capacity and good ductility, which is conducive to earthquake resistance.

The skeleton curves of different unbonded prestressed tendon pretension (tension control stress), different unbonded prestressed tendon area (number of roots), different slurry anchor bar unbonded length and different axial compression ratio are shown in Fig. 3.23.

Figure 3.23 shows that the shape and trend of the skeleton curve under various parameter changes are basically consistent. Among them, the increase of the axial compression ratio has a relatively obvious effect on the skeleton curve. As previously analyzed, the skeleton curve of specimen EHSW8 is basically consistent with that of specimen XJ at the initial stage of loading, and the two are gradually separated at the later stage.

Table 3.11 EHSW7 loading process

Loading process		Main phenomena
Load control	Elastic stage (0 ~ ± 300 kN)	The load and displacement change linearly; the loading and unloading curves are essentially coincident; and there is virtually no residual deformation after unloading
	Cracking stage (± 300 kN)	Horizontal cracks appear on the top surface of the specimen base and the upper layer of the slurry layer, with a maximum crack width of approximately 0.01 mm
	Yielding phase (± 500 kN)	The vertical reinforcement at the top edge of the tensile side of the wall yielded; the edge of the wall continued to produce new cracks along the height direction. The highest crack was about 1.5 m from the top surface of the base, and the higher the crack, the shorter the horizontal section, and the faster it developed downward to the neutral axis. Compared with Specimen XJ, the crack tilting occurred earlier and at a larger angle. In the range of about 50 cm height from the bottom of the specimen to the base, the positive and reverse oblique cracks had crossed the neutral axis and intersected to form cross oblique cracks. Similarly, obviously different from Specimen XJ, the cracks at the bottom of the specimen were less developed, and there were no cracks in the range of about 200 mm from the top of the base at both ends of the wall, with the existing cracks being close to the middle of the wall limb, which was the extension of the oblique cracks in the upper part of the wall During this stage, cracks continued to develop steadily, with a maximum width of approximately 0.24 mm. Joints also appear open and closed phenomenon, the maximum opening value is about 1.6 mm
Displacement control	Displacement loading stage	After entering the displacement loading stage, the wall continues to develop new cracks along the height direction. The highest crack is approximately 2.4 m from the top surface of the base and continues to extend obliquely downward. The positive and reverse cracks intersect over the neutral axis, and within a height range of about 1.4 m from the top surface of the base, positive and reverse cross-oblique cracks have formed When loaded to ± 46 mm stage, vertical cracks began appearing in the concrete protective layer. When loaded to the ± 69 mm stage, the protective layer separated from the core concrete, and the concrete of the local protective layer peeled off, while the exposed core concrete remained basically intact Loading to ± 69 mm stage, the maximum crack width increased from 0.7 mm to 3.0 mm; abutment opening is stable growth. Loading to ± 92 mm stage, abutment maximum opening value is about 19 mm

(continued)

Table 3.11 (continued)

Loading process		Main phenomena
	Destruction phase (± 92 mm)	When loaded to the second cycle of this stage, the outermost tensile steel bar of the specimen was broken individually, the exposed bellows were torn, the compressive side steel bar was obviously convex and buckling, the protective layer concrete was peeled off in a large range, the internal vertical steel bar and stirrup were exposed, and the core concrete was partially crushed The damage to the concrete is concentrated at both ends of the wall limb, with a damage range of approximately 10–20 cm from the top of the base in the height direction. The damage is less severe than that observed in specimen XJ

(a) wall side

(b) wall front

Fig. 3.16 Local failure of EHSW7 specimen

3.6 Bearing Capacity Analysis

3.6.1 Characteristic Point Bearing Capacity Analysis

Table 3.13 lists the cracking, yielding, and peak load of each specimen. Compared with specimen XJ, the load of the characteristic points of each EHSW specimen is higher. This shows that the initial pre-tension from the unbonded prestressed tendon effectively improves the crack resistance of the specimen and delays cracking. Additionally, although the unbonded tendon is located in the center of the wall, it still contributes to the bearing capacity of the specimen, resulting in a certain improvement in the yield and peak load of the specimen.

Figure 3.24 shows the bearing capacity of the characteristic points for specimens with different unbonded prestressed tendon pretension (tension control stress), different unbonded prestressed tendon area (number), different unbonded length of slurry anchor reinforcement, and different axial compression ratio.

Figure 3.24 shows that under the assumption that other parameters remain constant:

(1) With the increase of unbonded prestressed tendon pre-tension, it significantly affects the cracking load, causing it to increase linearly. However, this has no significant effect on the yield load or peak load. Analysis shows that the increase of pre-tension effectively delays specimen cracking and improves crack resistance. Nevertheless, because it has no effect on reinforcement ratio or prestressed strength ratio, it barely impacts the yield load or peak load of the specimen.
(2) With the increase of unbonded prestressed tendon area, specimen yield and peak loads increase, but cracking load remains largely unchanged. Analysis indicates that larger tendon area raises reinforcement ratio, enhancing yield and peak loads. However, as tendons are centrally located, their contribution to bearing capacity is limited, resulting in a relatively small increase (maximum

Table 3.12 EHSW8 loading process

Loading process		Main phenomena
Load control	Elastic stage (0 ~ ± 250 kN)	The load and displacement change linearly; the loading and unloading curves essentially coincide, and there is virtually no residual deformation after unloading
	Cracking stage (± 250 kN)	Horizontal cracks appear on the top surface of the specimen base and the upper layer of the slurry layer, with a maximum crack width of approximately 0.01 mm
	Yielding phase (± 525 kN)	The vertical reinforcement at the top edge of the tensile side of the wall has yielded; the edge of the wall continues to produce new cracks along the height direction. The highest crack is about 1.9 m from the top surface of the base, and the higher the crack, the shorter the horizontal section, and the faster it develops downward to the neutral axis. The shape of the crack is similar to that of specimen XJ, and the crack has an obvious horizontal section and tilts laterally. At the bottom of the specimen, the positive and reverse diagonal cracks have crossed the neutral axis within the height range of about 110 cm from the base, forming cross diagonal cracks. Similarly, obviously different from specimen XJ, the cracks at the bottom of the specimen are less developed, and there are no cracks in the range of about 260 mm from the top of the base at both ends of the wall, and the existing cracks are close to the middle of the wall limb, which is the extension of the oblique cracks in the upper part of the wall In this stage, crack development continues steadily, with a maximum crack width of approximately 0.42 mm. Joints also appear open and closed phenomenon, the maximum opening value of about 1.6 mm
Displacement control	Displacement loading stage	Upon entering the displacement loading stage, the wall continues to develop new cracks along the height direction; the highest crack is approximately 2.4 m from the top surface of the base. Based on the existing cracks, they continue to extend obliquely downward. Positive and reverse cracks intersect over the neutral axis, and within a height range of about 1.4 m from the top surface of the base, both positive and reverse cross-oblique cracks have been formed When loaded to ± 46 mm stage, vertical cracks began appearing in the concrete protective layer. When loaded to the ± 69 mm stage, the protective layer separated from the core concrete, and the concrete of the local protective layer peeled off; meanwhile, the exposed core concrete remained basically intact Loading to ± 69 mm stage, the maximum crack width increased from 0.7 mm to 2.0 mm; abutment opening is stable growth. Loading to ± 92 mm stage, the abutment maximum opening value is about 22 mm

(continued)

Table 3.12 (continued)

Loading process		Main phenomena
	Destruction phase (± 92 mm)	When loaded to the third cycle of this stage, the outermost tensile steel bar of the specimen was broken individually, exposing the bellows, which was torn; the compressive side steel bar was obviously convex and buckled; the protective layer concrete was peeled off in a large range, exposing the internal vertical steel bar and stirrups; and the core concrete was partially crushed The damage to the concrete is concentrated at both ends of the wall limb, with a damage range of approximately 10– 20 cm from the top of the base in the height direction. The damage is less severe than that observed in specimen XJ

(a) wall side (b) wall front

Fig. 3.17 Local damage of specimen EHSW8

< 10%) in cracking load; since total pretension remains constant, tendon area changes cannot significantly affect cracking load [10].

(3) With the increase of unbonded length of grouted bar, the yield load of the specimen increases gradually, the peak load decreases gradually, and the cracking load has no obvious effect. Analysis shows that the increase of unbonded length of grout-anchored steel bar can avoid strain concentration of steel bar to some extent and improve stress of steel bar, thus delaying yield of steel bar and increasing yield load of the specimen. At the same time, under the same level of displacement loading, the overstrength of steel bar will decrease with the increase of unbonded length, resulting in peak load decreasing with the increase of unbonded length [7].
(4) With increasing axial compression ratio, the cracking load, yield load, and peak load of specimens all increased progressively. The analysis indicates that the influence of axial compression ratio on the bearing capacity of EHSW specimens aligns with the mechanical properties of typical compression-bending members, namely that an increased axial compression ratio corresponds to an increased load at each stage of the specimen.

3.6.2 Performance Analysis of Bearing Capacity Reduction

The bearing capacity reduction performance can directly reflect the test specimen's ability to maintain load-bearing capacity. Higher values indicate slower and less premature bearing capacity decrease, thereby ensuring the specimen retains sufficient bearing capacity to avoid sudden collapse. For the bearing capacity reduction performance of specimens, the Specification for seismic test of buildings (JGJ/T 101-2015) [5] stipulates that the load reduction factor λ_i obtained by each cycle of the same level of loading is compared, and the calculation method is shown in Formula 3.5.

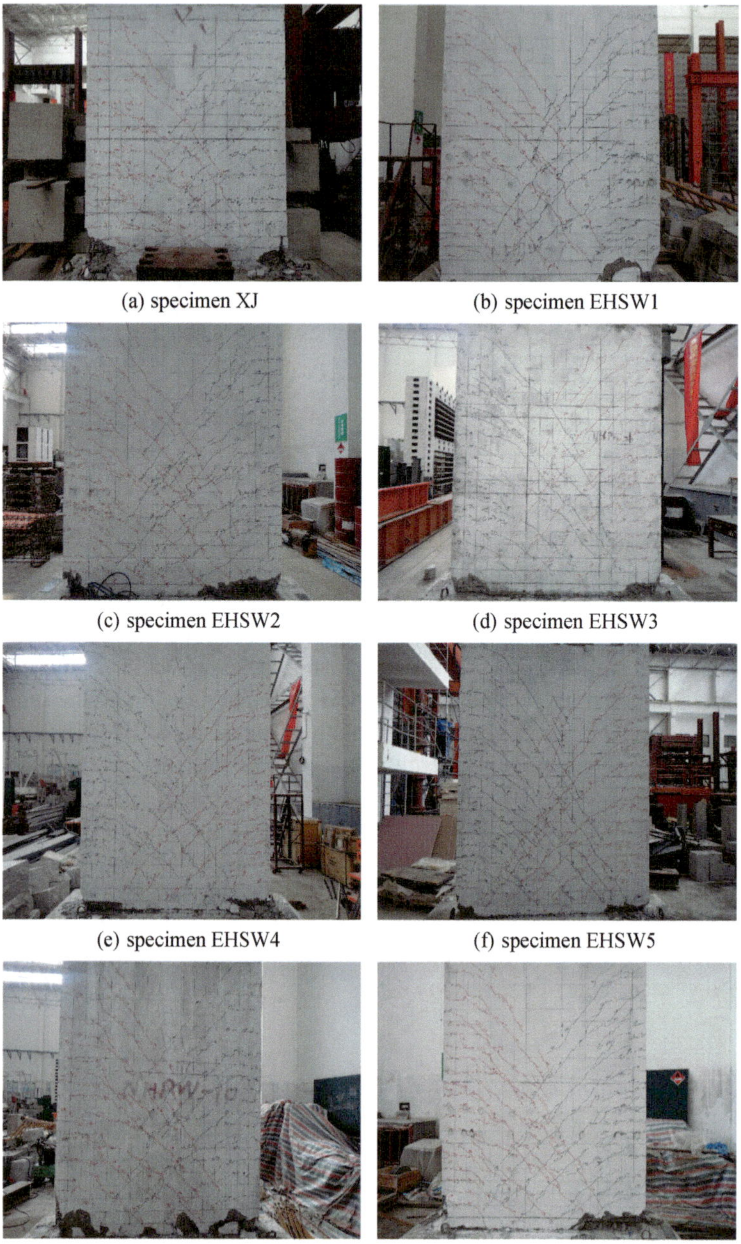

(a) specimen XJ
(b) specimen EHSW1
(c) specimen EHSW2
(d) specimen EHSW3
(e) specimen EHSW4
(f) specimen EHSW5
(g) specimen EHSW6
(h) specimen EHSW7

Fig. 3.18 Photos of final crack distribution of specimen

(i) specimen EHSW8

Fig. 3.18 (continued)

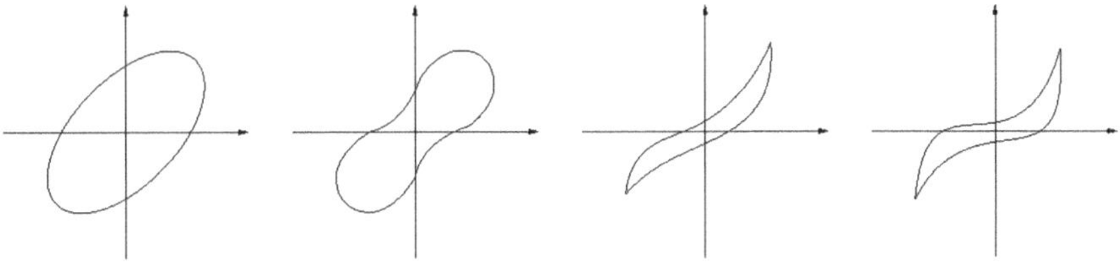

Fig. 3.19 Shape of hysteresis curve

$$\lambda_i = \frac{F_j^i}{F_j^{i-1}} \tag{3.5}$$

F_j^i when the displacement ductility factor is j, the *i*th cycle peak load value;
F_j^{i-1} when the displacement ductility factor is j, the i-1 cycle peak load value.

In this test, the peak load value of each cycle under the same stage loading displacement is calculated. The load reduction coefficient for the first cycle of each stage displacement is 1. The calculation results for the load reduction coefficient of each specimen during the displacement loading process are summarized in Table 3.14. In the table, "loading process" is a column. "Δ" represents the unit loading displacement, that is, 23 mm. The number in front of "Δ" represents the displacement amplitude, and the number behind "Δ" represents the number of cycles. "+" represents positive loading, and "–" represents negative loading. For instance, "-4Δ3" represents the third cycle of – 92 mm loading level. In addition, for some specimens that did not complete the final loading level of the entire loading cycle (e.g., specimen XJ only

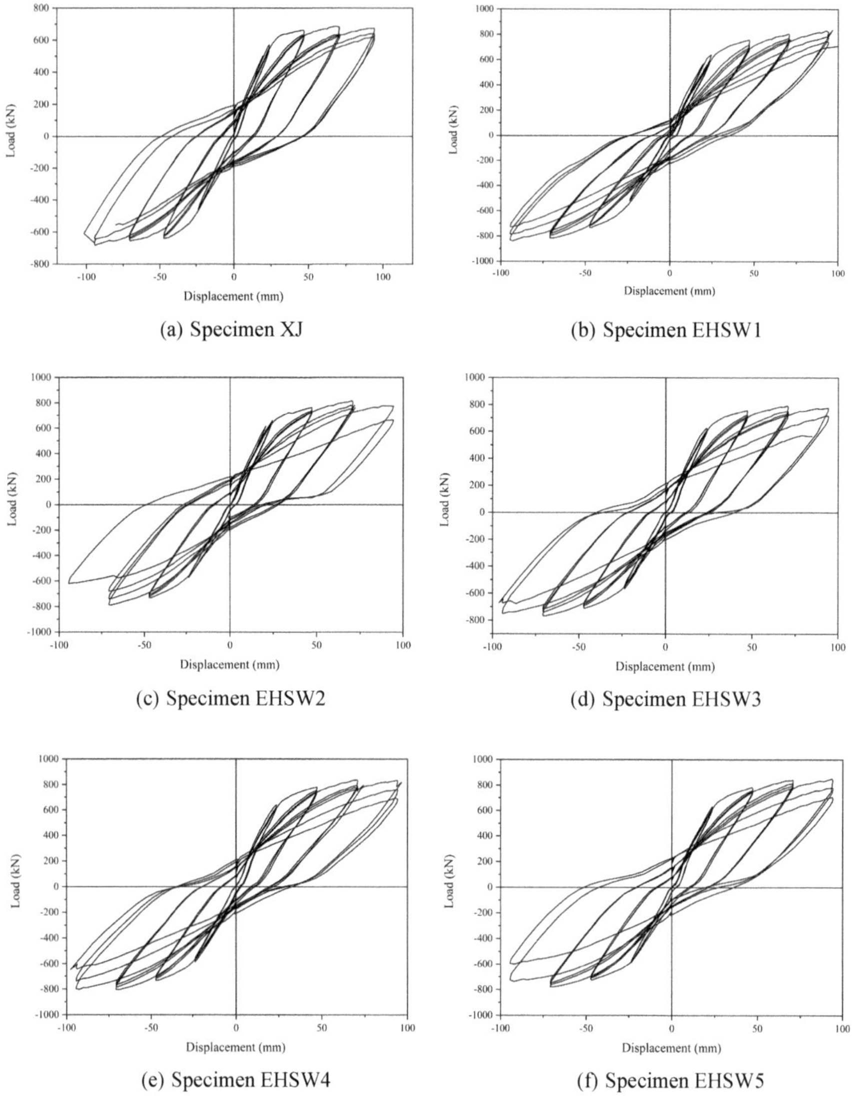

Fig. 3.20 Hysteretic curve of specimen

completed the 4th displacement of the 3rd cycle of positive loading and unloading process), the table is not done statistics.

As can be seen from Table 3.14, the load reduction coefficient of each specimen during the loading process is high, indicating that the bearing capacity of the specimen is reduced slowly and can be stably maintained under large deformation. In the elastic–plastic stage, it still exhibits good overall stability. The bearing capacity of each specimen is the same, that is, under the loading condition of 3 cycles per

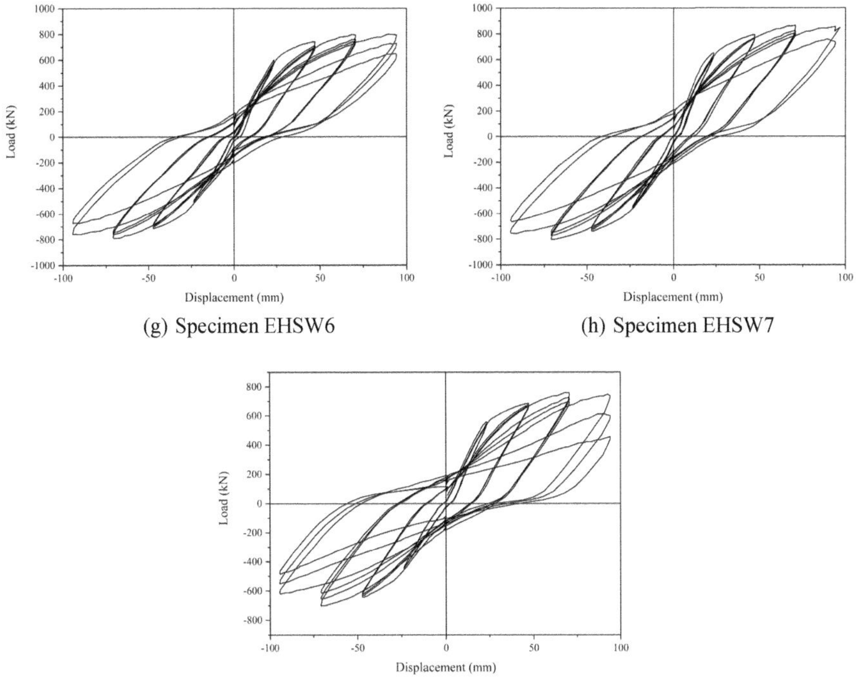

(g) Specimen EHSW6

(h) Specimen EHSW7

(i) Specimen EHSW8

Fig. 3.20 (continued)

displacement, the bearing capacity of the second cycle decreases more, while the bearing capacity of the third cycle can basically maintain the bearing capacity of the second cycle.

3.7 Stiffness Degradation Analysis

According to the Specification for seismic test of buildings (JGJ/T 101-2015) [5], the stiffness of the specimen can be expressed by secant stiffness, and the calculation method of secant stiffness K_i is shown in Formula 3.6.

$$K_i = \frac{|+F_i| + |-F_i|}{|+X_i| + |-X_i|} \tag{3.6}$$

F_i the ith cycle peak load value
X_i the ith cycle peak displacement value

The process of changing secant stiffness in each cycle for the elastic stage, yield stage (first yield of vertical reinforcement in shear wall), and displacement loading

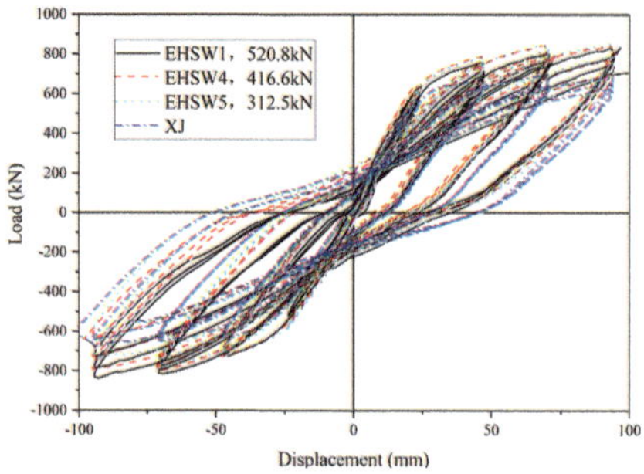

(a) Pretension of different unbonded prestressing tendons

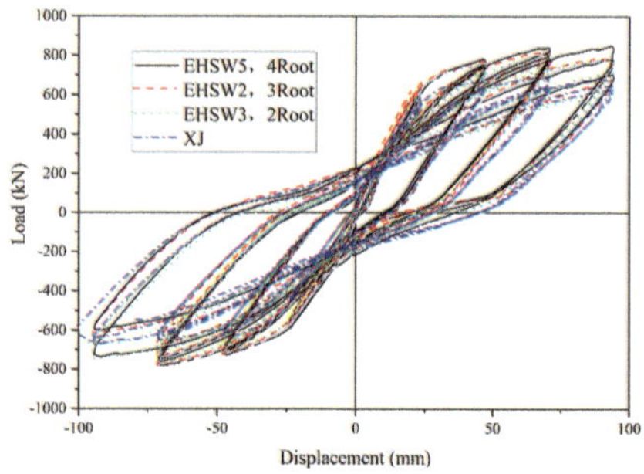

(b) Different unbonded prestressed tendon area

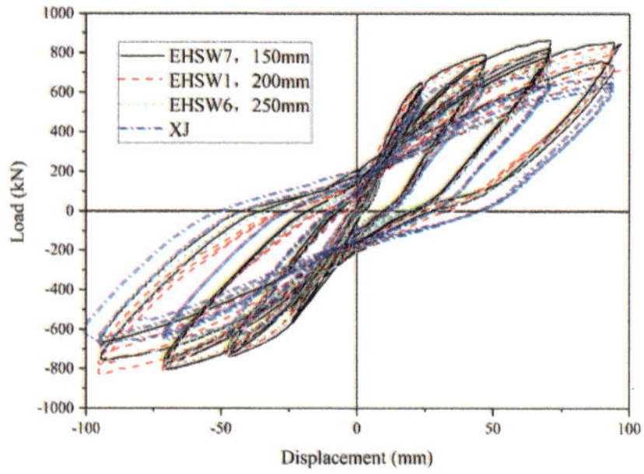

(c) Unbonded length of different grouted anchor bars

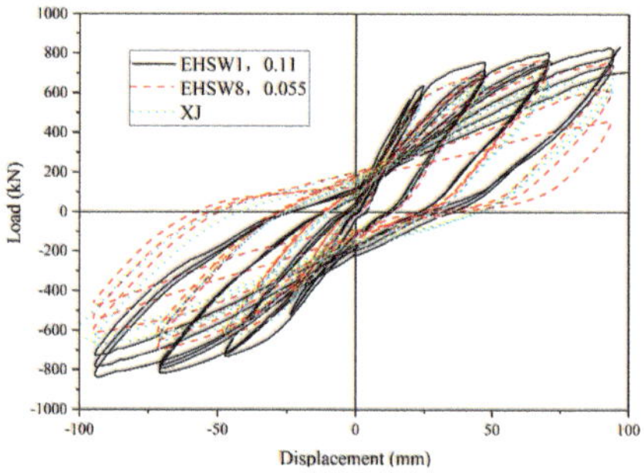

(d) Different axial compression ratios

Fig. 3.21 Comparison of specimen hysteresis curves

stage is shown in Fig. 3.25. The stiffness variation of each specimen is essentially the same. In different grades of cracking, yielding, and displacement loading, the stiffness significantly decreases. In each cycle of displacement loading at the same level, the stiffness also decreases to a certain extent, but the amplitude is relatively small. With the increase of loading level, the stiffness of the specimen gradually degrades, but the degradation rate gradually slows down, and the curve gradually flattens with the increase of displacement level.

Different unbonded prestressed tendon pretension (tension control stress), different unbonded prestressed tendon area (number), different slurry anchor steel unbonded length, and different axial compression ratio resulted in variations in the entire process of stiffness change, elastic stiffness, and yield stiffness. These changes are depicted in Fig. 3.26.

As can be preliminarily seen from Fig. 3.26, the elastic stiffness and yield stiffness of EHSW specimens with other parameter combinations are higher than those of XJ specimens to a certain extent, except that the decrease of axial compression ratio leads to the decrease of elastic stiffness and yield stiffness of EHSW specimens compared

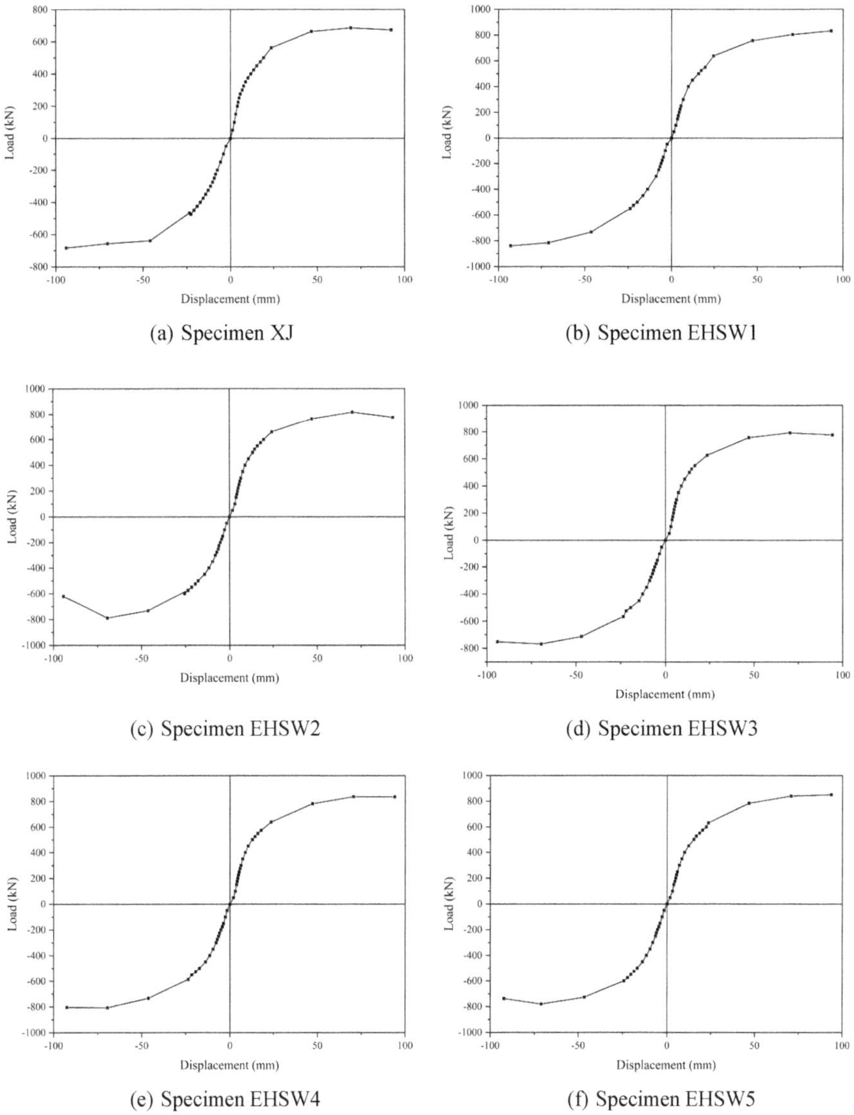

Fig. 3.22 Specimen skeleton curve

with XJ specimens. At the same time, under the premise that other parameters remain unchanged:

(1) With the increase of pre-tension of unbonded prestressed tendon, the elastic stiffness and yield stiffness of the specimen are gradually increased. The analysis shows that the change of pre-tension of unbonded prestressed tendon will directly affect the stiffness of the specimen, especially the elastic stiffness. With

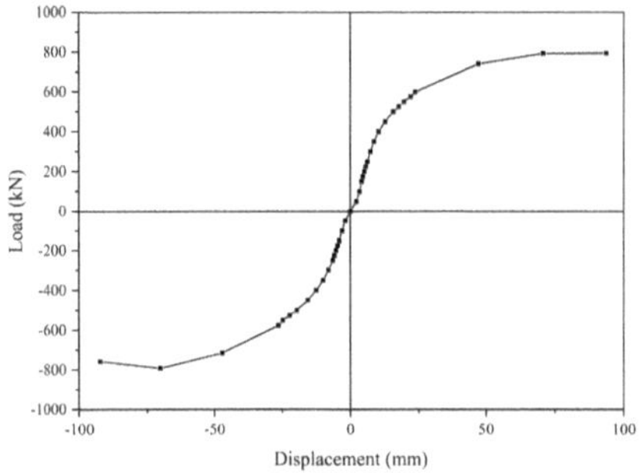

(g) Specimen EHSW6

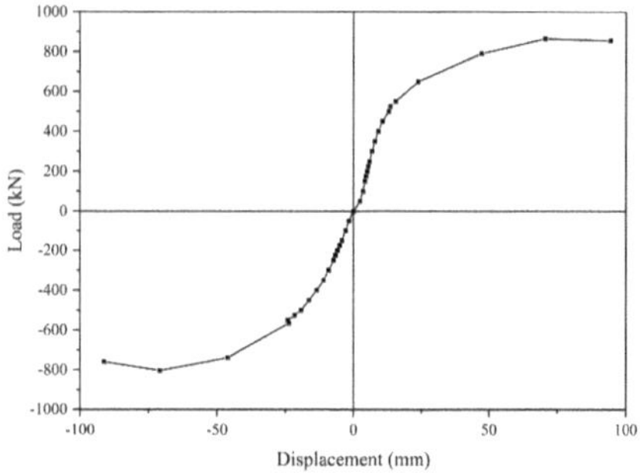

(h) Specimen EHSW7

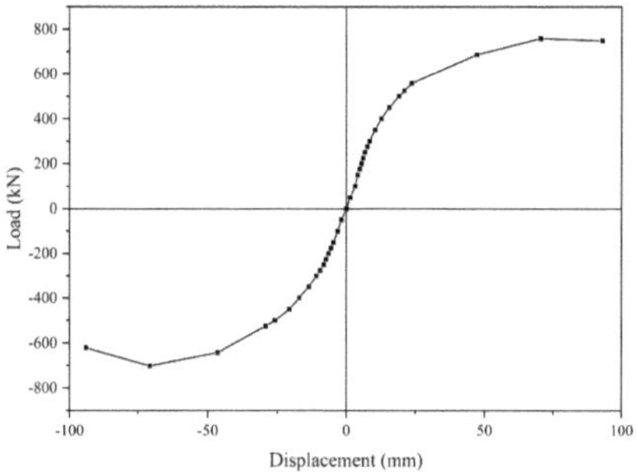

(i) Specimen EHSW8

Fig. 3.22 (continued)

the increase of pre-tension of prestressed tendon, the elastic stiffness and yield stiffness are improved, with the latter increasing at a faster rate.

(2) With the increase of unbonded prestressed tendon area, the elastic stiffness and yield stiffness of the specimen exhibit a non-monotonic variation, initially increasing before decreasing. The analysis shows that while the unbonded prestressed tendon area differs, its direct and obvious influence on the specimen's stiffness characteristics is minimal because the pre-tension in the unbonded prestressed tendon remains constant. As the prestressed tendon area increases, the stiffness of the specimen does not exhibit a clear regularity, and the change magnitude is small. The stiffness improvement effect is evident post-yield until the 2Δ stage, indicating that unbonded prestressed compression significantly improves stiffness before and after yield. With increasing displacement loading, the stiffness improvement effect gradually wanes [10].

(3) With the increase of the unbonded length of the grouted anchor bar, the elastic stiffness of the specimen initially increases and then decreases, while the yield stiffness always decreases. The analysis shows that the change of unbonded length of grout-anchored steel bar directly delays the yield of steel bar, and because of its unbonded structure, the overall elongation of steel bar increases when it yields, which increases the yield displacement of the specimen, i.e., the

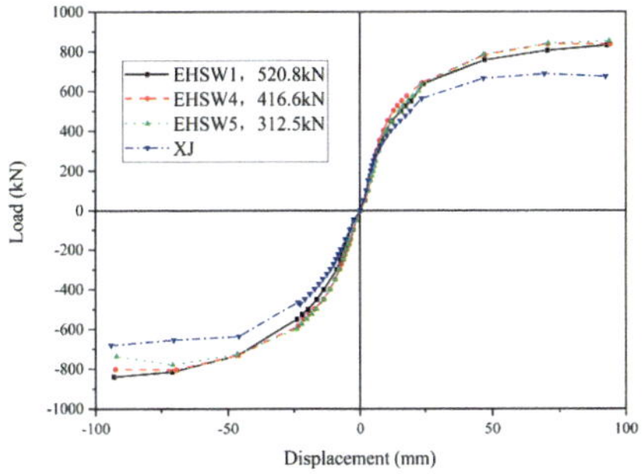

(a) Pretension of different unbonded prestressing tendons

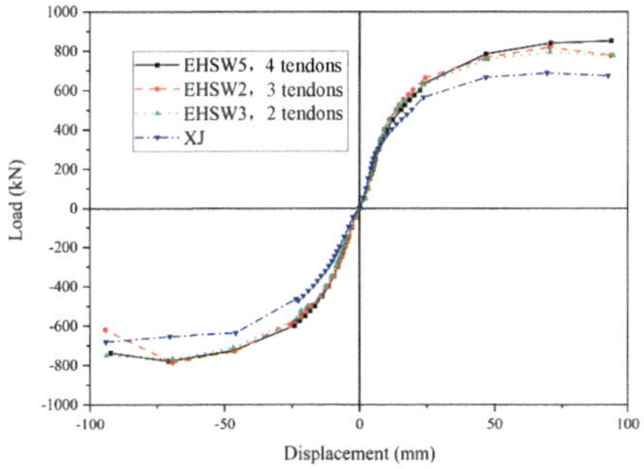

(b) Different unbonded prestressed tendon area

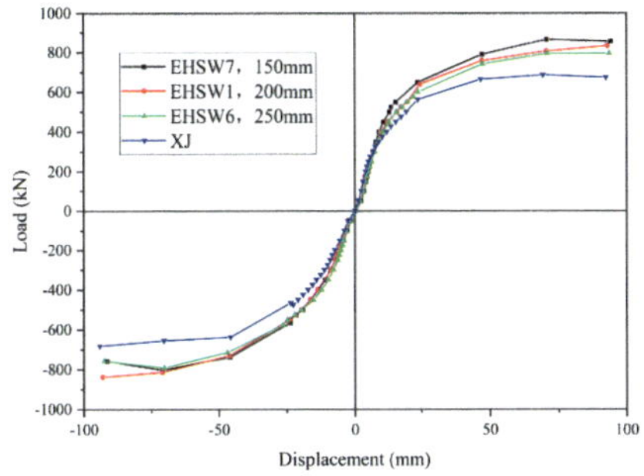

(c) Unbonded length of different grouted anchor bars

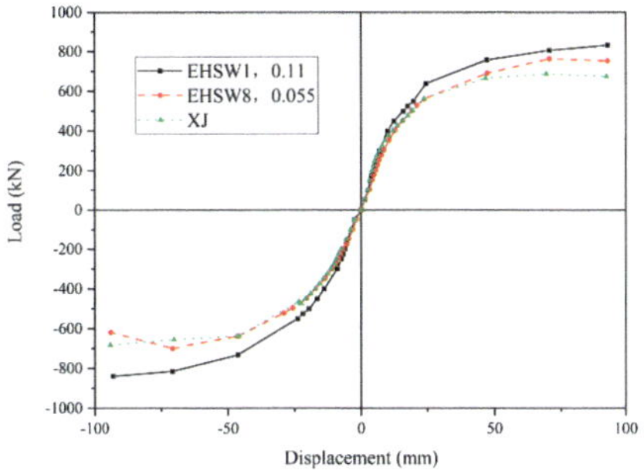

(d) Different axial compression ratios

Fig. 3.23 Comparison of skeleton curves of specimens

Table 3.13 The bearing capacity data of characteristic points of specimens

Specimen	Cracking load[a]/kN	Yield load[b]/kN	Peak load[c]/kN
XJ	225	475	685
EHSW1	300	550	839
EHSW2	250	525	816
EHSW3	250	500	792
EHSW4	275	550	836
EHSW5	250	550	851
EHSW6	300	575	796
EHSW7	300	500	864
EHSW8	225	500	759

Note [a] Cracking load is the horizontal load corresponding to the first crack of the specimen
[b] The yield load is the horizontal load corresponding to the first yield of the vertical reinforcement
[c] The peak load takes the maximum value of horizontal load in the whole loading process

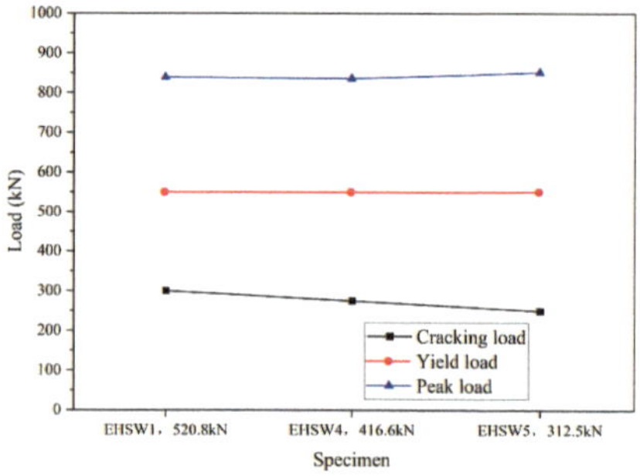

(a) Pretension of different unbonded prestressing tendons

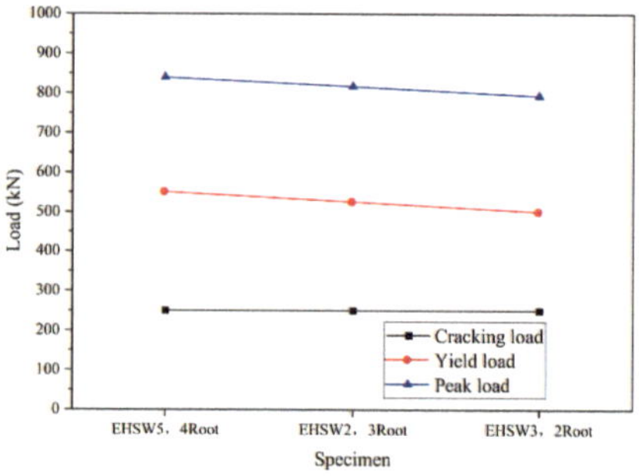

(b) Different unbonded prestressed tendon area

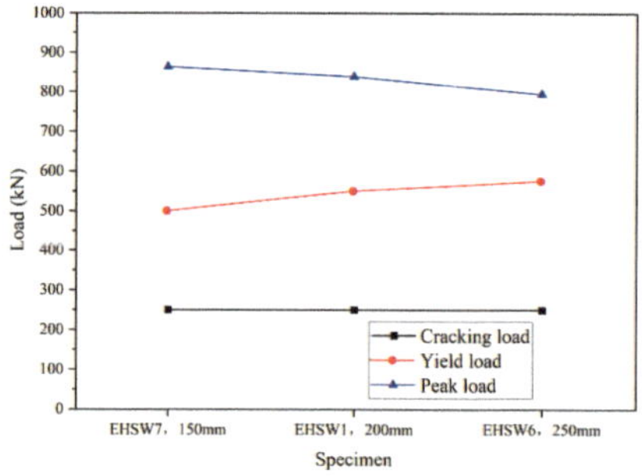

(c) Unbonded length of different grouted anchor bars

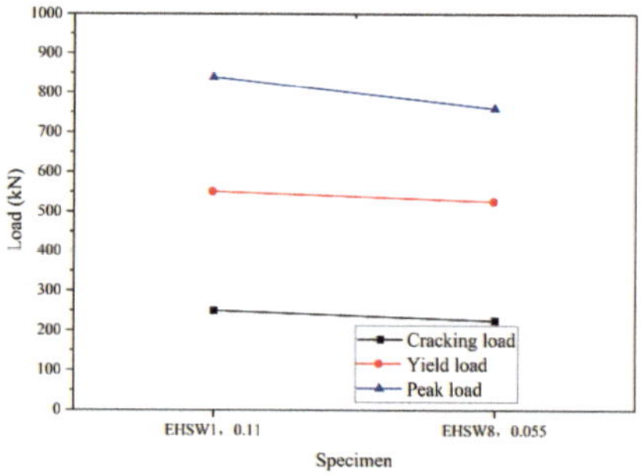

(d) Different axial compression ratios

Fig. 3.24 Comparison of bearing capacity at characteristic points of specimens

yield stiffness of the specimen decreases. However, due to the limited stress of the steel bar in the elastic stage, the strain is small, and the elongation difference caused by the unbonded length is not significant. Therefore, the unbonded length of the grouted steel bar has no direct effect on the elastic stiffness of the specimen. With the increase of the unbonded length, the elastic stiffness of the specimen exhibits no clear regularity, manifesting as the elastic stiffness of the data increasing first and then decreasing. Meanwhile, after the specimen yields until the 2Δ stage, due to the different unbonded lengths, the stiffness of the specimen is discrete. After entering the 2Δ stage, the influence of the unbonded length basically disappears, and each EHSW specimen exhibits very close stiffness characteristics [7].

(4) With increasing axial compression ratio, elastic stiffness and yield stiffness significantly improve. Analysis shows axial compression ratio's influence aligns with unbonded prestressed tendon pretension. As axial compression ratio decreases, elastic stiffness and yield stiffness decline rapidly. Under low axial compression ratio conditions, EHSW specimen stiffness is even lower than XJ specimen.

Table 3.14 Load reduction factor of specimen

Loading process	Test specimen								
	XJ	EHSW1	EHSW2	EHSW3	EHSW4	EHSW5	EHSW6	EHSW7	EHSW8
− 4Δ3	–	0.927	–	–	0.878	–	–	–	0.882
− 4Δ2	0.944	0.946	–	0.906	0.926	0.819	0.886	0.870	0.902
− 4Δ1	1.000	1.000	1.000	1.000	1.000	1.000	1.000	1.000	1.000
− 3Δ3	0.991	0.982	0.923	0.975	0.976	0.983	0.978	0.970	0.943
− 3Δ2	0.979	0.974	0.940	0.961	0.978	0.968	0.972	0.963	0.941
− 3Δ1	1.000	1.000	1.000	1.000	1.000	1.000	1.000	1.000	1.000
− 2Δ3	0.997	0.996	0.990	0.987	0.992	0.993	0.993	0.986	0.986
− 2Δ2	0.976	0.978	0.979	0.971	0.977	0.985	0.975	0.984	0.984
− 2Δ1	1.000	1.000	1.000	1.000	1.000	1.000	1.000	1.000	1.000
− 1Δ3	0.998	0.996	1.000	0.991	0.988	0.998	0.998	0.995	0.995
− 1Δ2	0.996	0.996	0.997	0.982	0.979	0.997	1.000	0.988	1.000
− 1Δ1	1.000	1.000	1.000	1.000	1.000	1.000	1.000	1.000	1.000
+ 1Δ1	1.000	1.000	1.000	1.000	1.000	1.000	1.000	1.000	1.000
+ 1Δ2	0.951	0.918	0.958	0.973	0.973	0.983	0.975	0.972	0.986
+ 1Δ3	0.996	0.990	0.995	0.990	0.989	0.992	0.988	0.987	0.993
+ 2Δ1	1.000	1.000	1.000	1.000	1.000	1.000	1.000	1.000	1.000
+ 2Δ2	0.963	0.948	0.975	0.954	0.963	0.964	0.958	0.962	0.977
+ 2Δ3	0.997	0.974	0.990	0.988	0.983	1.000	0.983	1.000	0.996
+ 3Δ1	1.000	1.000	1.000	1.000	1.000	1.000	1.000	1.000	1.000
+ 3Δ2	0.921	0.943	0.957	0.963	0.949	0.967	0.956	0.953	0.949
+ 3Δ3	0.992	0.997	0.976	0.973	1.000	0.969	0.976	0.973	0.965
+ 4Δ1	1.000	1.000	1.000	1.000	1.000	1.000	1.000	1.000	1.000
+ 4Δ2	0.952	0.921	–	0.921	0.928	0.917	0.919	0.896	0.816
+ 4Δ3	–	0.884	–	–	0.904	–	–	–	0.730

3.8 Displacement Ductility Analysis

Ductility refers to the deformation capacity of a structure, component or a section of a component from the beginning of yielding to the maximum bearing capacity or after reaching the bearing capacity, without significantly reducing its strength. The late deformation capacity of a ductile structure, component or section is large, and it can absorb a certain amount of energy after reaching the yield or maximum bearing capacity state, which can avoid brittle failure. Therefore, ductility is an important parameter reflecting the plastic deformation capacity of structures or components, and is also one of the important indicators measuring the seismic performance of structures. Structural seismic design, especially for high-rise buildings, is often

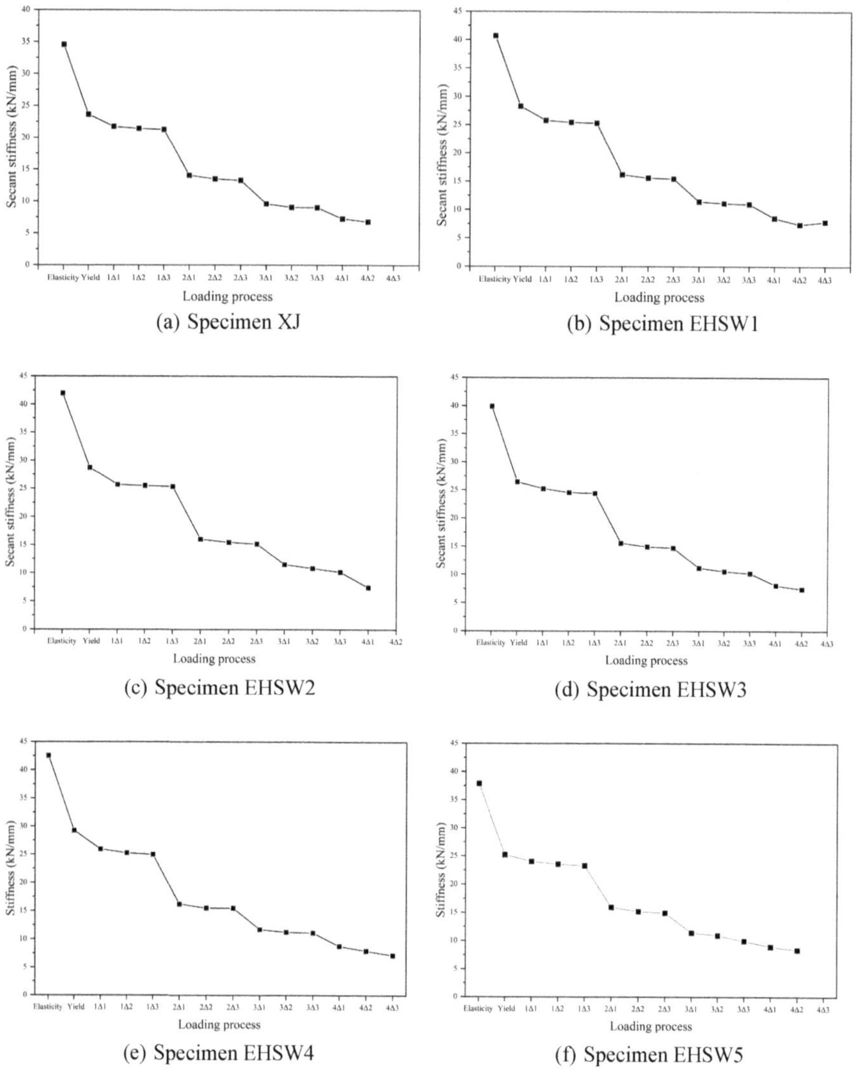

Fig. 3.25 Specimen stiffness degradation curve

based on ductile design at this stage, aiming for good ductility to enable structures or components to accommodate accidental overload or repeated load generated by internal forces and deformation. Post-ductile structures have large deformation capacity, which can serve as a structural safety reserve.

The ductility performance of structures or components is typically characterized by ductility coefficients, including curvature, rotation, and displacement. These coefficients reflect ductility characteristics at different levels, such as section, component,

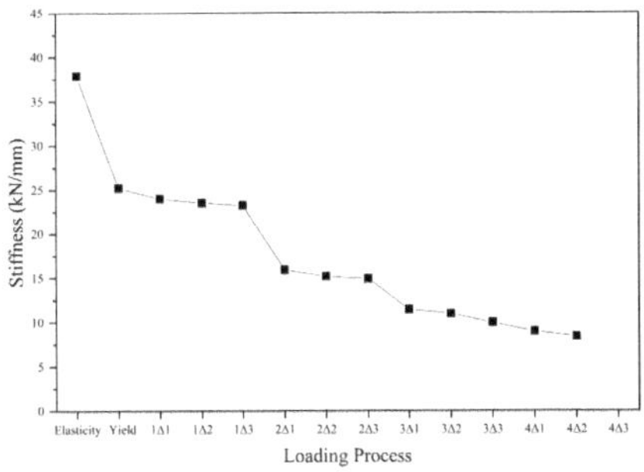

(g) Specimen EHSW6

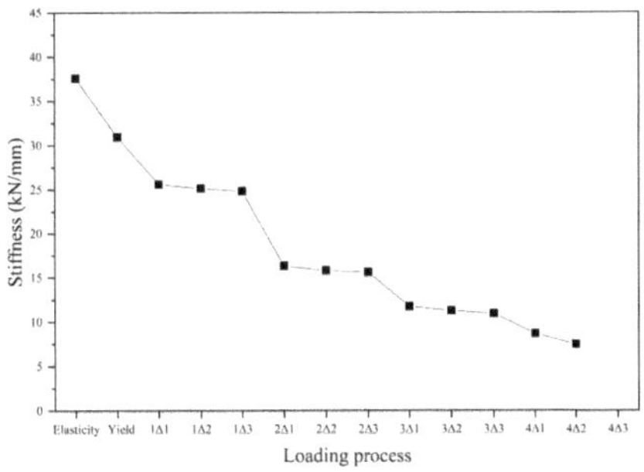

(h) Specimen EHSW7

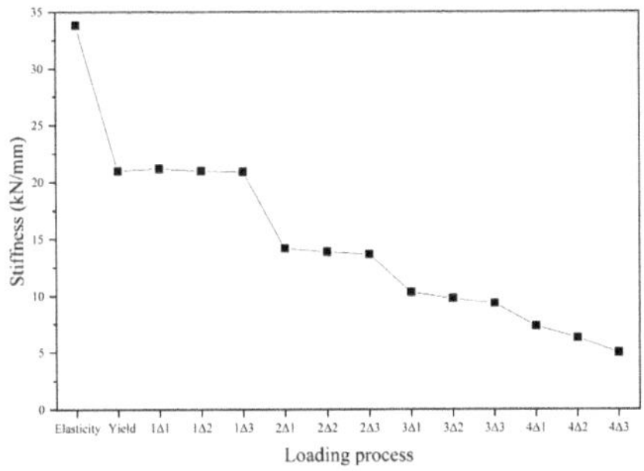

(i) Specimen EHSW8

Fig. 3.25 (continued)

or structure. This book uses the displacement ductility factor to analyze specimen ductility characteristics.

To calculate the displacement ductility factor, refer to the Specification for seismic test of buildings (JGJ/T 101-2015) [5], and use the calculation method shown in Eq. 3.7.

$$\mu = \frac{\Delta_u}{\Delta_y} \tag{3.7}$$

Δ_u limit displacement of specimen;
Δ_y the yield displacement of the specimen.

For the determination of the ultimate displacement Δ_u, the displacement value corresponding to the load falling to 85% of the peak load can be taken according to the provisions of the Code for Seismic Test Methods of Buildings (JGJ/T 101-2015) [5]. Since the load of most specimens in this book did not decrease significantly until the end of the test, the end displacement of the test was taken as the ultimate displacement.

There are many methods for determining the yield displacement Δ_y, including modified general bending moment yield method, energy equivalent method, and

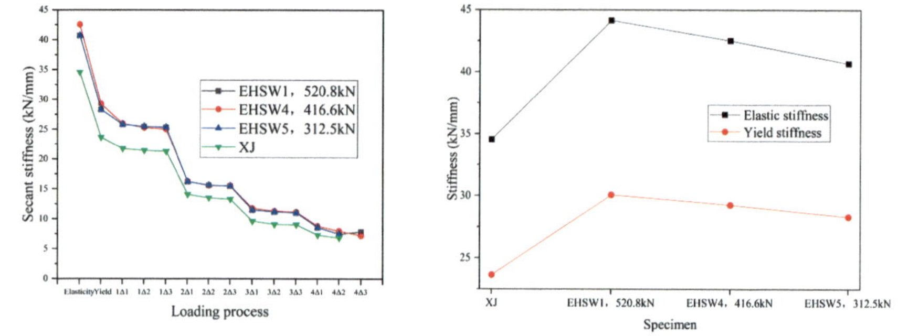

(a) Pretension of different unbonded prestressing tendons

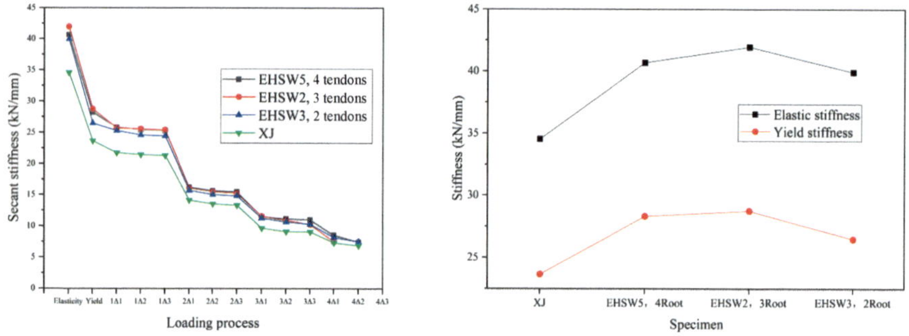

(b) Different unbonded prestressed tendon area

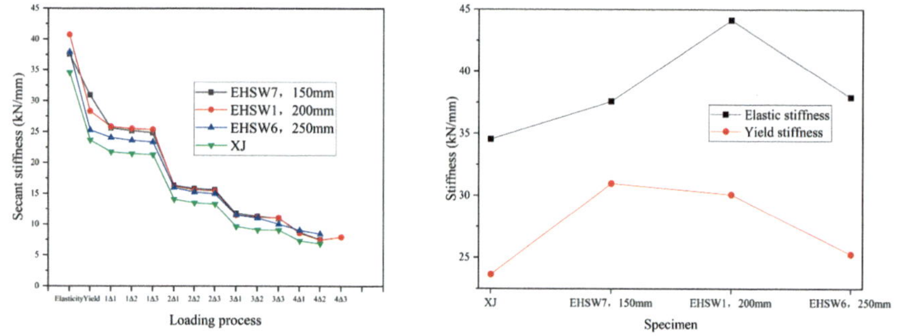

(c) Unbonded length of different grouted anchor bars

Fig. 3.26 Comparison of specimen stiffness

geometric drawing method. The Park method is employed in this book, and the determination of yield displacement is relatively straightforward, in good agreement with previous test data. The basic process of determining yield displacement is as follows: in the skeleton curve of the specimen, a horizontal line is drawn through the peak load point, and then another horizontal line is drawn through 75% of the peak load to intersect with the skeleton curve at a point. A secant line is drawn through

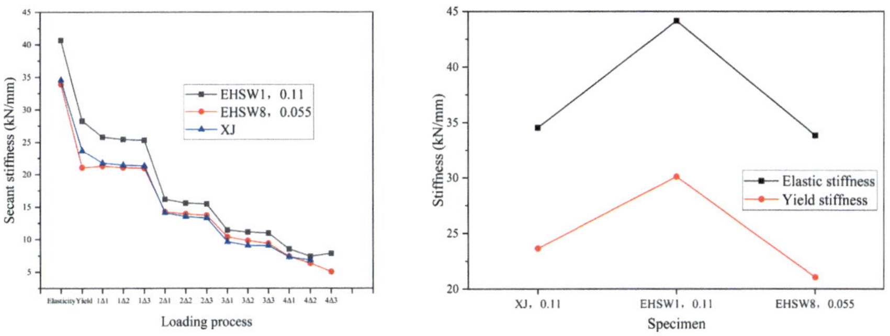

(d) Different axial compression ratios

Fig. 3.26 (continued)

this point and the origin, intersecting the horizontal line passing through the peak load at one point. A vertical line is then drawn downward from this point to intersect the skeleton curve at another point. This second point is the yield point, and the displacement value of this point is the yield displacement value to be determined [11]. The solving process is depicted in Fig. 3.27.

The yield displacement, ultimate displacement, and calculated displacement ductility coefficient of each specimen are listed in Table 3.15. As can be seen from the table, except for EHSW1 and EHSW6, the average displacement ductility coefficient of EHSW specimens is close to or even higher than that of XJ, indicating that EHSW specimens have good ductility performance and are conducive to earthquake resistance.

By comparing the average displacement ductility coefficient of contrast specimens with different unbonded prestressing tendons (tension control stress), different unbonded prestressing tendon areas (number), different unbonded lengths of slurry

Fig. 3.27 Yield displacement determined by Park method

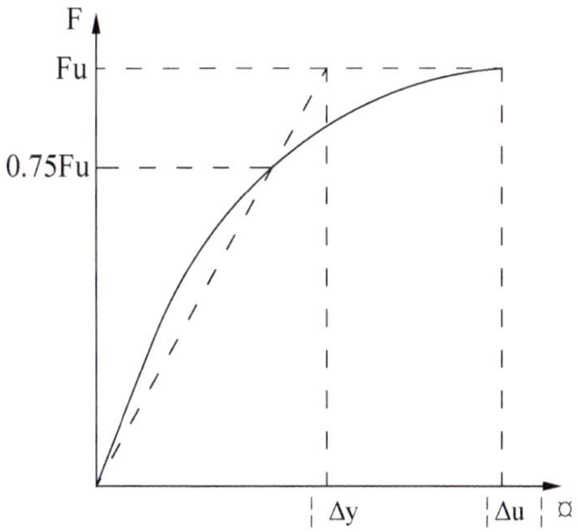

Table 3.15 Specimen displacement ductility coefficient

Specimen	Yield displacement/mm		Ultimate displacement/mm		Displacement ductility factor		Average displacement ductility factor
	Positive direction	Negative direction	Positive direction	Negative direction	Positive direction	Negative direction	
XJ	27.3	– 39.7	92.4	– 94.2	3.38	2.37	2.875
EHSW1	31.2	– 44.5	92.9	– 93.1	2.98	2.09	2.535
EHSW2	28.5	– 34.1	93.3	– 94.3	3.27	2.77	3.020
EHSW3	27.5	– 32.9	94.2	– 94.0	3.43	2.86	3.145
EHSW4	30	– 35.6	94.1	– 92.5	3.14	2.60	2.870
EHSW5	32.4	– 30	92.2	– 92.2	2.85	3.07	2.960
EHSW6	31.6	– 44.2	93.5	– 92.5	2.96	2.09	2.525
EHSW7	29.1	– 34.8	94.3	– 91.2	3.24	2.62	2.930
EHSW8	28.4	– 36.7	92.9	– 93.9	3.27	2.56	2.915

anchor steel bars, and different axial compression ratio parameters, it can be seen that under the premise of other parameters remaining unchanged:

(1) Increasing the pre-tension of unbonded prestressed tendon will reduce the displacement ductility factor. The analysis shows that increasing the pre-tension of the unbonded prestressed tendon increases the initial compressive strain of the vertical steel bar, delays the yield of the steel bar, and increases the yield displacement of the specimen, thus reducing the displacement ductility of the specimen.
(2) A reduction in unbonded prestressed tendon area can improve the displacement ductility of the specimen. Analysis shows that the area change can directly alter the reinforcement ratio and prestressed strength ratio of the specimen. Therefore, when the area is decreased, the reinforcement ratio and prestressed strength ratio are reduced, thereby enhancing the ductility of the specimen. This is also consistent with the provisions of the "Standard for seismic design of prestressed concrete structures (JGJ/T 140-2019)" [12] that limit the amount of prestressed tendons.
(3) An increase in the unbonded length of grouted anchor steel bars can reduce the displacement ductility coefficient of specimens. Analysis indicates that augmenting the unbonded length of anchor bars can delay steel yield, increase yield displacement, and decrease displacement ductility of specimens.
(4) A decrease in axial compression ratio can improve the displacement ductility of specimens. Analysis indicates that the influence of axial compression ratio on displacement ductility is identical to that of unbonded prestressed tendon pre-tension effect. Consequently, a reduction in axial compression ratio significantly enhances specimen displacement ductility.

3.9 Energy Dissipation Capacity Analysis

The structure is subjected to seismic energy during an earthquake, and the earthquake inputs energy into the structure. The structure undergoes a continuous process of energy absorption and dissipation. When the structure enters the elastoplastic state, its seismic performance mainly depends on the energy dissipation capacity of the structure. Under cyclic loading, the structure is expected to have sufficient energy dissipation capacity without significant loss of bearing capacity and stiffness. The area surrounded by the load–displacement curve in the loading stage of the hysteresis curve can reflect the amount of energy absorbed by the structure, while the shadow area surrounded by the curve in the unloading stage and the loading curve is the dissipated energy. This energy is converted into heat and lost to space via internal friction or local material damage. Therefore, the area of the hysteresis loop in the load–displacement hysteresis curve is an important index for evaluating energy dissipation in a structure.

There are many indexes to characterize the energy dissipation capacity of structures or components. The equivalent viscous damping coefficient h_e and the working index I_W are usually used. In this paper, the equivalent viscous damping coefficient h_e is used to determine the energy dissipation capacity of the test specimen. The calculation method is shown in Fig. 3.28 and Eq. 3.8.

$$h_e = \frac{1}{2\pi} \cdot \frac{S_{ABCD}}{S_{\Delta OBE} + S_{\Delta ODF}} \tag{3.8}$$

S_{ABCD} the area of hysteresis loop ABCD;
S_{OBE} the area of triangular OBE;
S_{ODF} the area of triangle ODF.

The equivalent viscous damping coefficients of each cycle of each specimen at each loading level in the displacement loading stage are calculated and listed in Table 3.16. According to the values in the table, while the energy dissipation performance of EHSW specimens varies, some values are greater than specimen XJ, some are less, but overall the data are not substantially different. The maximum difference between any two is within 20% (in the initial stage of displacement loading, i.e., 1Δ1 loading stage, the difference ratio between specimen EHSW5 and specimen XJ). As the loading level increases, the difference decreases (still taking specimen EHSW5 as an example, until the 4Δ2 loading stage, the difference ratio between the two decreases to less than 8%). Therefore, it can be considered that the EHSW specimens have good energy dissipation capacity.

The equivalent viscous damping coefficients for specimens with different pretension of unbonded prestressed tendons (tension control stress), different area of unbonded prestressed tendons (number of roots), different unbonded length of grouted steel bars, and different axial compression ratio parameters are depicted in Fig. 3.29.

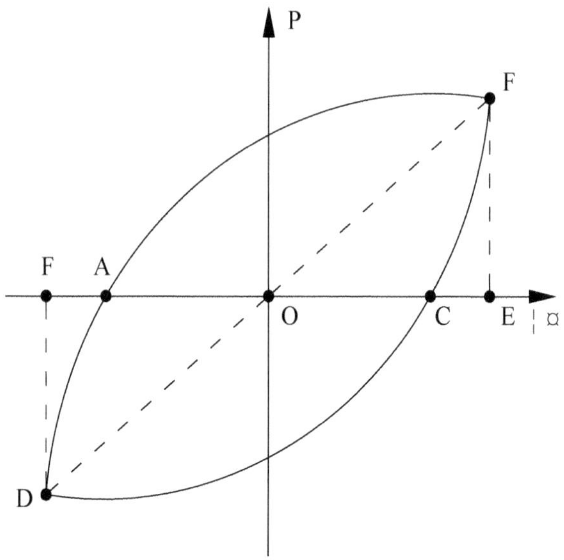

Fig. 3.28 Calculation of equivalent viscous damping coefficient

Table 3.16 Equivalent viscous damping coefficient of specimen

Load process	Test specimen								
	XJ	EHSW1	EHSW2	EHSW3	EHSW4	EHSW5	EHSW6	EHSW7	EHSW8
1Δ1	0.01825	0.02195	0.01997	0.02007	0.02006	0.01468	0.01608	0.02156	0.02154
1Δ2	0.01356	0.01499	0.01457	0.01479	0.01487	0.01215	0.01311	0.01428	0.01688
1Δ3	0.01246	0.0136	0.0137	0.01396	0.01403	0.01164	0.01224	0.01328	0.01593
2Δ1	0.05948	0.05343	0.06032	0.06229	0.05694	0.05952	0.05102	0.05584	0.06047
2Δ2	0.04459	0.0369	0.04676	0.04811	0.04255	0.04509	0.03264	0.04232	0.04867
2Δ3	0.04104	0.03323	0.04342	0.04528	0.03891	0.04267	0.03013	0.03947	0.04626
3Δ1	0.07661	0.05902	0.07242	0.07512	0.06602	0.06976	0.04788	0.07939	0.07565
3Δ2	0.06932	0.05245	0.06794	0.07212	0.05929	0.06426	0.03963	0.07027	0.06976
3Δ3	0.0672	0.05039	0.06592	0.07114	0.06035	0.06249	0.03704	0.06966	0.06811
4Δ1	0.09398	0.06954	0.08625	0.08893	0.07787	0.08481	0.05836	0.08881	0.08427
4Δ2	0.08413	0.06465	–	0.07905	0.07561	0.07752	0.05232	0.08711	0.08329
4Δ3	–	0.06175	–	–	0.07259	–	–	–	0.08240

It can be seen from Fig. 3.29 that under the premise that other parameters remain unchanged:

(1) The pre-tension of unbonded prestressed tendon increases, and the energy dissipation capacity of the specimen decreases. Analysis shows that the increase of pre-tension of unbonded prestressed tendons will delay the yield of steel bars. Under the same ultimate displacement condition, the yield degree of slurry

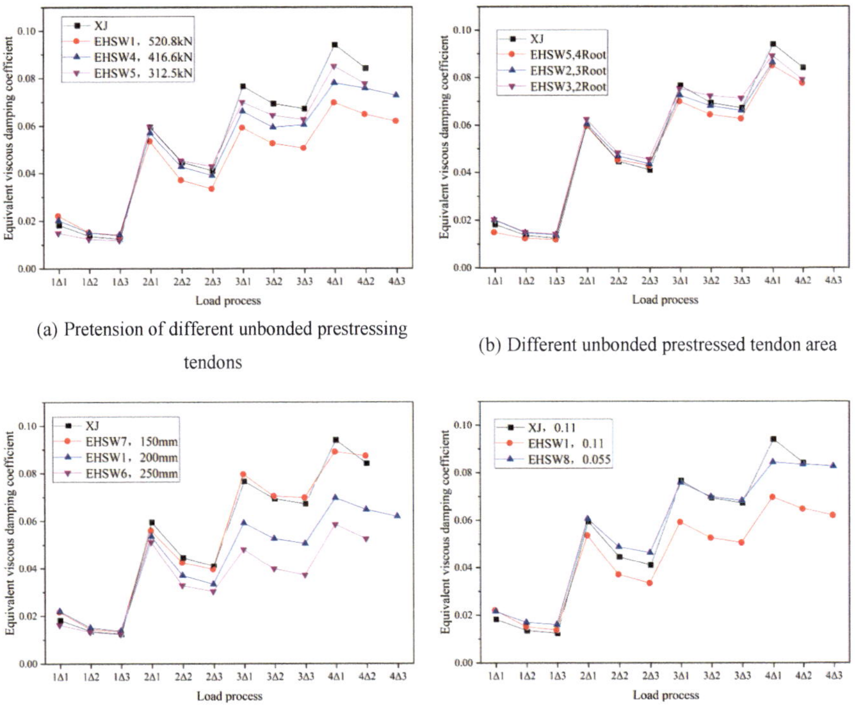

(a) Pretension of different unbonded prestressing tendons

(b) Different unbonded prestressed tendon area

(c) Unbonded length of different grouted anchor bars

(d) Different axial compression ratios

Fig. 3.29 Comparison of energy dissipation capacity of specimens

anchor steel bars will be reduced, thereby limiting the yield energy consumption of steel bars and reducing the energy consumption capacity of specimens to a certain extent.

(2) With the increase of unbonded prestressed tendon area, the equivalent viscous damping coefficient and energy dissipation capacity of specimens decrease. The energy dissipation capacity of EHSW specimens with this parameter is close to that of XJ specimens. The analysis shows that the prestressed strength ratio of the specimen increases with the increase of unbonded prestressed tendon area, while the proportion of ordinary steel bars bearing the energy dissipation mechanism is relatively reduced, which will weaken the energy dissipation capacity of the specimen.

(3) The energy dissipation capacity of the specimen decreases significantly with the increase in the unbonded length of the grouted steel bars. The analysis indicates that increasing the unbonded length delays the yield of the steel bars, reduces their yield level, and limits the energy dissipation during yielding, thus lowering the specimen's overall energy dissipation capacity. In the later stage of the test, energy consumption is primarily attributed to the yielding of the ordinary

steel bars. Among the specimens, EHSW7, with the shortest unbonded length, exhibits stress changes in the steel bars that are closest to those in specimen XJ, thus demonstrating slightly better performance than the other EHSW specimens during the later stage of loading [7].

(4) The energy dissipation capacity of the EHSW8 specimen with a small axial compression ratio is basically consistent with that of the XJ specimen. The analysis shows that the influence of deformation recovery ability on the overall performance of the specimen is more prominent under larger displacement loading grades due to the introduction of unbonded prestressed tendons into the EHSW specimen, manifested as hysteresis loop pinching during unloading. Therefore, under the same conditions, the energy dissipation capacity of EHSW1 is weaker than that of XJ. The effect of axial compression ratio on the energy dissipation capacity of the specimen is akin to the influence of the elastic restoring force of unbonded prestressed tendons. The vertical load also serves to return the specimen to its initial position during specimen unloading. Therefore, a reduction in the axial compression ratio will also decrease the energy dissipation capacity of the specimen, consistent with the energy dissipation characteristics of traditional compression-bending members. This can effectively enhance the energy dissipation capacity of the specimen [6].

3.10 Residual Deformation Analysis

The residual deformation of a structure characterizes its deformation recovery ability, and this ability directly affects the structure's performance, repairability, and repair cost after an earthquake.

Figure 3.30 plots the residual deformation of each specimen at the yield stage and subsequent displacement loading stage. The positive and negative loads of each loading cycle are unloaded to zero. It can be seen that the residual deformation of each specimen has basically the same development law. In the initial stage of loading (i.e., the yield stage and the first loading stage, ± 23 mm), the residual deformation of the specimen is small. After entering the second loading stage (± 46 mm), the residual deformation increases significantly. In addition, except for the fourth stage of loading (± 92 mm), which may be close to the failure of the specimen, the residual deformation varies greatly and the regularity is inconsistent. The residual deformation of each cycle of loading at other stages does not change much, indicating that the residual deformation is stable during each stage of displacement loading.

Figure 3.31 shows that the residual deformation of each EHSW specimen is smaller than that of specimen XJ, with other parameters unchanged.

(1) The pre-tension of unbonded prestressed tendon increases, and the residual deformation of the specimen decreases obviously. The analysis shows that the increase of the pre-tension of the unbonded prestressed tendon directly causes the elastic restoring force of the prestressed tendon to increase synchronously

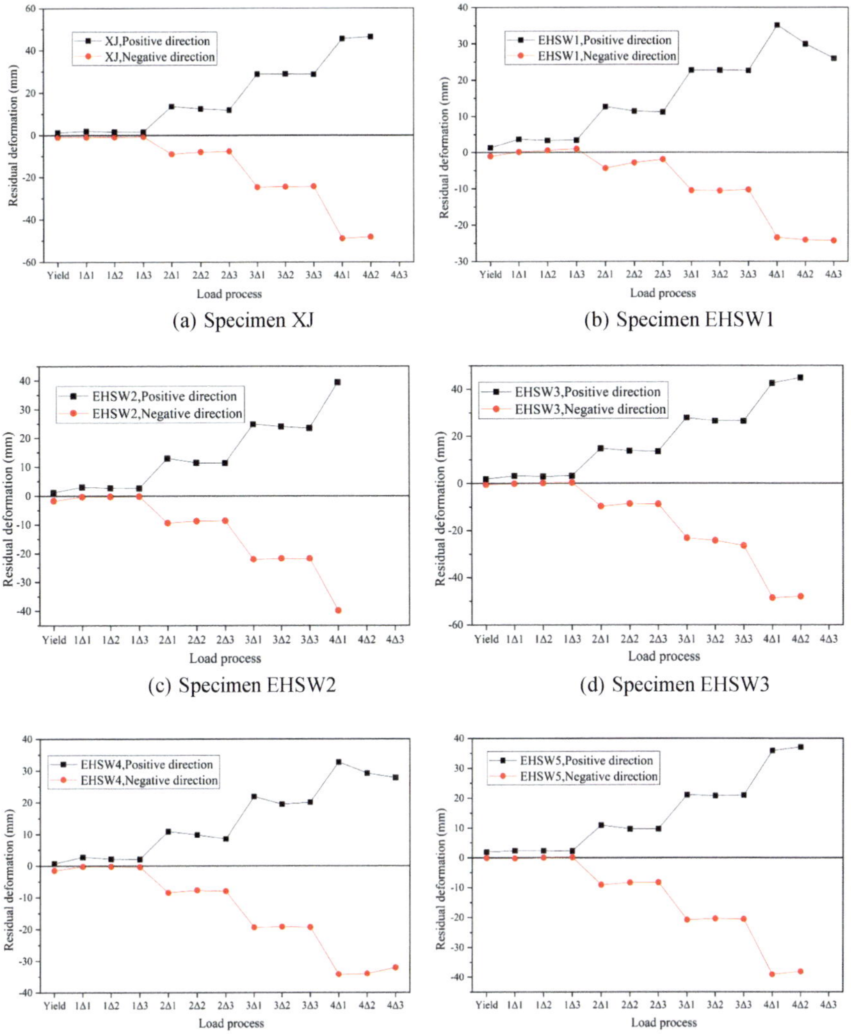

(a) Specimen XJ
(b) Specimen EHSW1
(c) Specimen EHSW2
(d) Specimen EHSW3
(e) Specimen EHSW4
(f) Specimen EHSW5

Fig. 3.30 Residual deformation curve of specimen

under all levels of load, thus significantly reducing the residual deformation of the specimen.

(2) With the increase of unbonded prestressed tendon area, the specimen's residual deformation decreases, but the degree is not significant. The analysis shows that the change in unbonded prestressed tendon area has little effect on the elastic restoring force provided by the prestressed tendon, which determines the residual deformation of the specimen. Therefore, although the measured residual

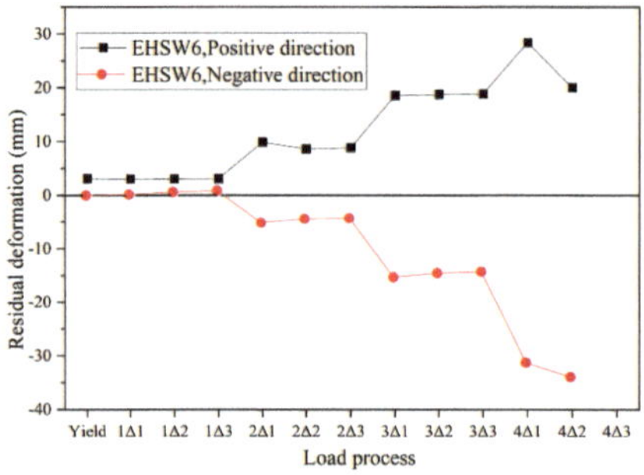

(g) Specimen EHSW6

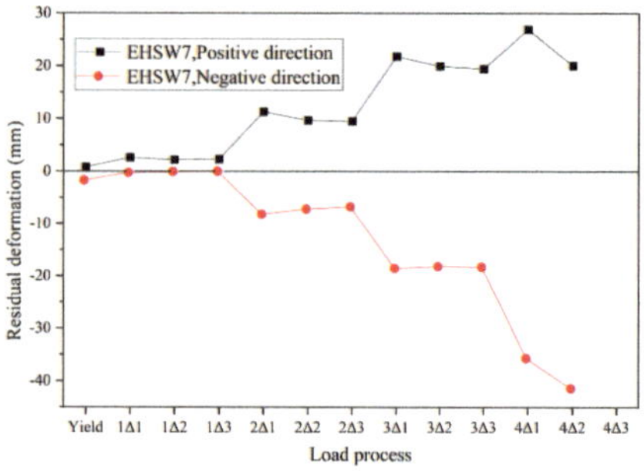

(h) Specimen EHSW7

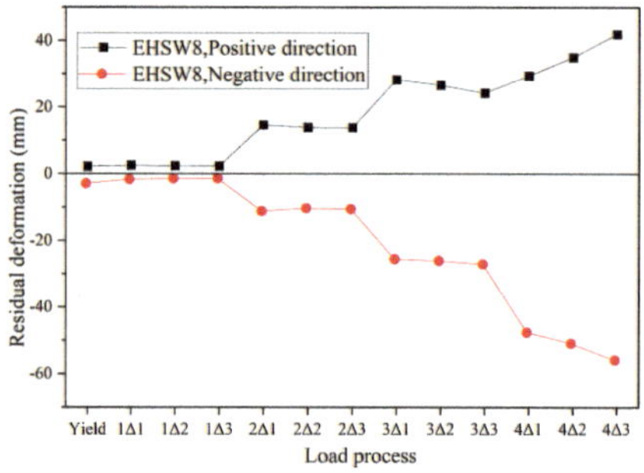

(i) Specimen EHSW8

Fig. 3.30 (continued)

deformation of the specimen decreases with the increase of the prestressed tendon area, the amplitude is not significant enough.

(3) When the unbonded length of grouted steel bar increases, the residual deformation of the specimen exhibits no obvious pattern. Analysis indicates that the grouted anchor reinforcement lacks bond and cannot directly influence the stress and deformation of the unbonded prestressed reinforcement. This has no effect on the elastic recovery force of the prestressed reinforcement and no direct relationship with the residual deformation of the specimen.

(4) The axial compression ratio also significantly influences the residual deformation of the specimen. A decrease in the axial compression ratio will obviously increase the residual deformation of the specimen. Analysis reveals that the influence of the axial compression ratio on residual deformation is analogous to the pre-tension effect of unbonded prestressed tendons [6], which can be viewed as a fixed part of the elastic restoring force provided by unbonded prestressed tendons. Therefore, with an increase in the axial compression ratio, the residual deformation of the specimen decreases.

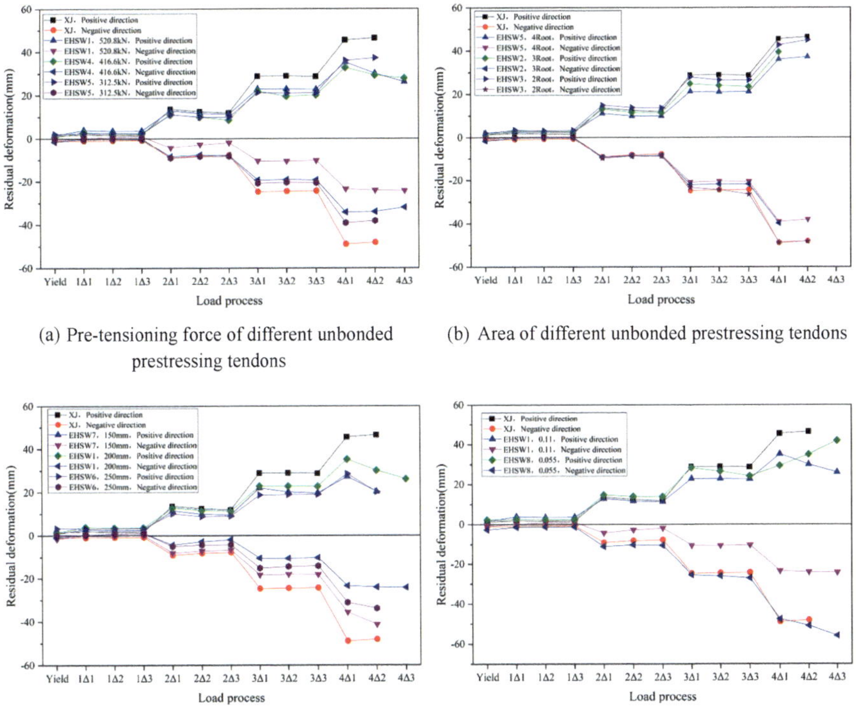

(a) Pre-tensioning force of different unbonded prestressing tendons

(b) Area of different unbonded prestressing tendons

(c) Unbonded length of different slurry-anchored steel bars

(d) Different axial pressure ratios

Fig. 3.31 Comparison of specimen residual deformation

3.11 Stress/Strain Analysis of Rebar

In the test, embedded anchor cable pressure sensors are used to measure the tension changes of the unbonded prestressed tendon during loading, while strain gauges attached to the slurry-anchored steel bar and stirrup monitor the strain variations in the steel bars. By analyzing the stress and strain development of the steel bars, the stress state of the specimen can be intuitively understood, enabling further analysis of macroscopic mechanical characteristics, such as yield behavior and energy dissipation. The book will provide the measured stress value of prestressed tendons based on the measured pressure sensor results and the theoretical cross-sectional area of the prestressed tendons. For steel strain gauges, due to their limited range (tensile and compressive strain limits of 0.02), and potential for data distortion leading to data drift and overflow from strain gauge slippage or breakage at later stages of loading, some singular data are properly processed.

In addition, due to the large number of strain gauges, in order to save space and facilitate analysis, only the strain values of key parts are given.

3.11.1 Stress Analysis of Prestressed Tendon

The stress/tension state of unbonded prestressed tendons throughout the loading process is critically important for the stress and deformation of EHSW specimens. As discussed above, the stress state of the prestressed tendon directly determines the elastic recovery force it can provide, which will directly influence the deformation recovery ability of the EHSW specimen. Moreover, the stress of prestressed tendons will directly determine the bearing capacity of the specimen, affecting its ultimate strength.

The stress of prestressed tendons in the initial stage and post-axial compression stage of each EHSW specimen, as well as the peak stress and stress increment during loading, are listed in Table 3.17. Due to the limitations of the prestressing tendon tensioning process and inherent defects such as gap and slip in tensioning tooling and equipment, the initial prestressing value of the tendon differs from the predetermined value, but the difference is small and basically meets the requirements of the parameter change needed for the test.

Table 3.17 provides a simple method to calculate the stress change in prestressed reinforcement after axial compression. This can be done by using the theoretical value from Formula 3.9. From this formula, it can be observed that, under the given specimen conditions, the stress change in this stage is solely related to the applied axial pressure and exhibits a positive proportional relationship with it. The stress change is not directly proportional to the number of prestressed reinforcement.

$$\Delta\sigma = E_s \cdot \frac{N}{E_c A_w} \tag{3.9}$$

E_s Prestressed steel strand elastic modulus, take 1.95×10^5 MPa,
E_c C30 modulus of elasticity for concrete, take 3.15×10^4 MPa,

Table 3.17 EHSW Unbonded prestressed tendon stress of specimen

Specimen	Number of prestressed tendons	Initial prestress/MPa	Prestress after axial compression/MPa	Stress change value after axial compression/MPa	Peak stress/MPa	Stress increment/MPa
EHSW1	4	880.3	864.2	– 16.1	1158.9	278.6
EHSW2	3	758.5	741.8	– 16.7	1082.1	323.6
EHSW3	2	1266.1	1248.2	– 17.9	1416.1	150.0
EHSW4	4	676.8	658.9	– 17.9	948.2	271.4
EHSW5	4	530.3	514.2	– 16.1	808.9	278.6
EHSW6	4	827.6	811.5	– 16.1	1127.6	300.0
EHSW7	4	893.7	877.6	– 16.1	1177.6	283.9
EHSW8	4	885.2	877.1	– 8.10	1167.4	282.1

A_w Section area of specimen wall, calculated as 340000 mm^2,
N Vertical force, includes two parts. One part of the specimen, in addition to the base of other parts of the weight, takes a theoretical value of 32.9 kN. The other part is the external axial pressure. According to the different axial pressure ratio, it can be taken as 946.3 kN, 473.15 kN.

Therefore, according to Formula 3.9, the prestress loss of EHSW1~7 prestressed tendons after axial compression is 17.8 MPa, and the corresponding loss of EHSW8 is 9.2 MPa. Comparing the theoretical value with the measured value, it can be found that the measured value and theoretical value exhibit consistent regularity. Although the theoretical value does not consider factors such as material parameter dispersion and equipment process error, there is good agreement between the theoretical and measured values, which also confirms the reliability of the measured value from another perspective.

Additionally, it was observed that the peak stress of the prestressed tendons in each EHSW specimen does not exceed their yield strength (1650 MPa, see Table 3.3), ensuring that the unbonded prestressed tendons can provide reliable and stable elastic restoring force throughout the test. This enhances the specimen's self-recovery ability. However, once the yield strength is exceeded, plastic deformation of the prestressed tendons significantly affects the restoring force, making the self-recovery ability unreliable and unusable.

3.11.2 Strain Analysis of Ordinary Steel Bar

The strain gauges on the outermost tensile vertical steel bars and the bottom horizontal stirrups (about 80 mm from the base) near the joints (about 30 mm from the base) of each test specimen were chosen to record data, which was plotted in Fig. 3.32.

Due to the local unbonded structure of the slurry anchor steel bar in the EHSW specimen, the strain concentration of the steel bar is weakened, the steel bar strain is reduced, and yield of the slurry anchor steel bar occurs at a higher load level. The yield load of the EHSW specimen in Table 3.13 is higher than that of the XJ specimen and the lag of the yield strain of the steel bar in Fig. 3.32. Meanwhile, for comparison of yield load of EHSW specimens with different unbonded lengths, the longer the unbonded length, the higher the weakening degree of stress concentration, thus delaying steel yield further. However, it should be noted that, as mentioned above, increasing unbonded length is unfavorable to energy dissipation and ductility of the specimen.

For the vertical steel strain, on one hand, due to the Bauschinger effect of the steel material, and on the other hand, after the failure of the concrete protective layer, the steel bar is subjected to repeated tension and compression yield, resulting in some steel bars bending and twisting in the opposite direction. This is shown in Fig. 3.32, which illustrates that some steel bar strains are essentially located above the X axis during later stages of loading.

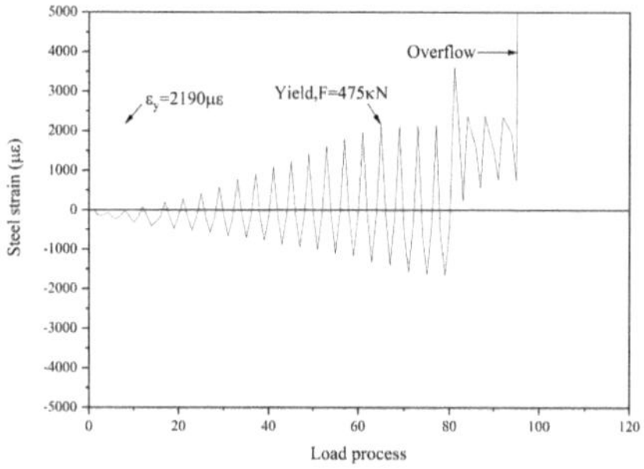

(a) Specimen XJ vertical reinforcement strain

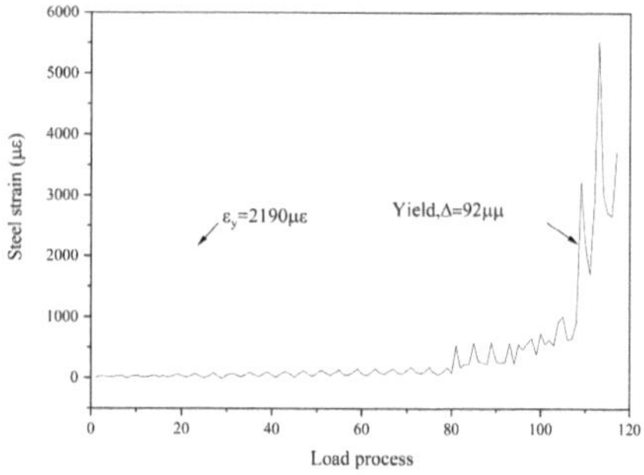

(b) Specimen XJ stirrup strain

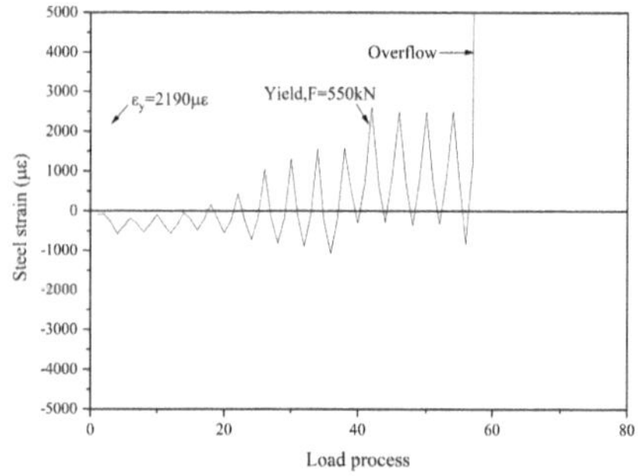

(c) Specimen EHSW1 vertical reinforcement strain

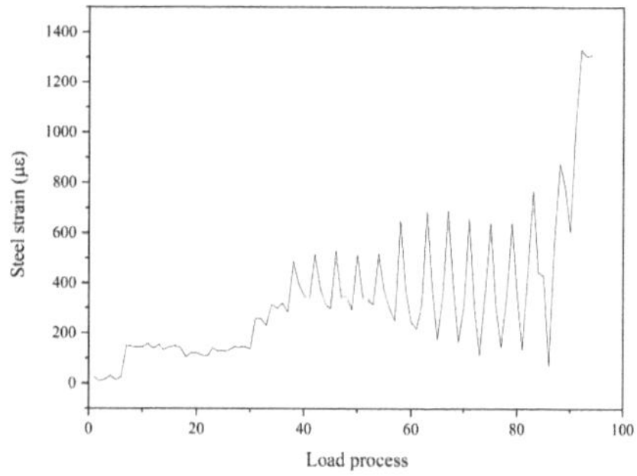

(d) Specimen EHSW1 stirrup strain

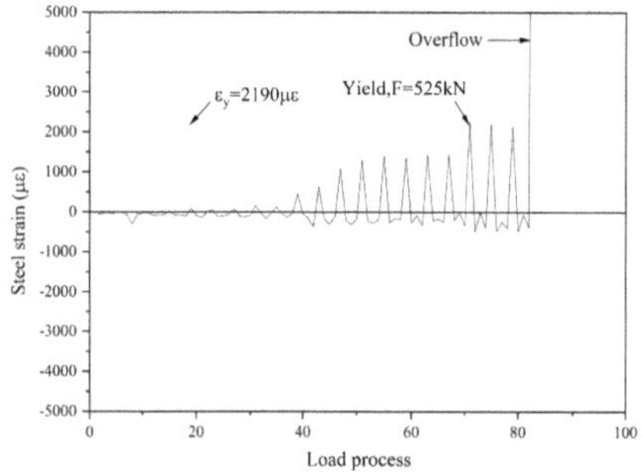

(e) Specimen EHSW2 vertical reinforcement strain

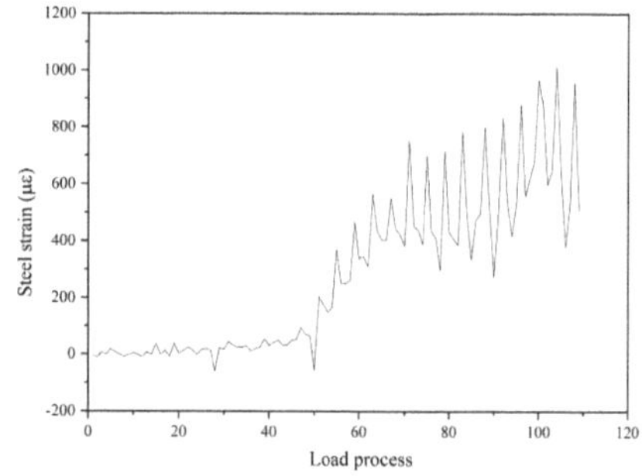

(f) Specimen EHSW2 stirrup strain

Fig. 3.32 Specimen ordinary steel strain

For the stirrup strain, the stirrup is essentially in a tensile state. Additionally, it can be observed that the EHSW specimen is more continuous and reliable due to the use of the buckle closed stirrup structure. Consequently, when the specimen is destroyed, all the stirrups of the specimen have not yet yielded and still provide good constraint on the concrete. By improving the ductility of the concrete material, the damage

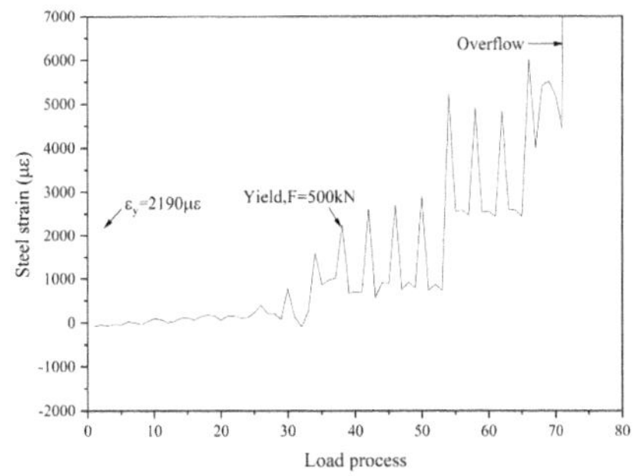

(g) Specimen EHSW3 vertical reinforcement strain

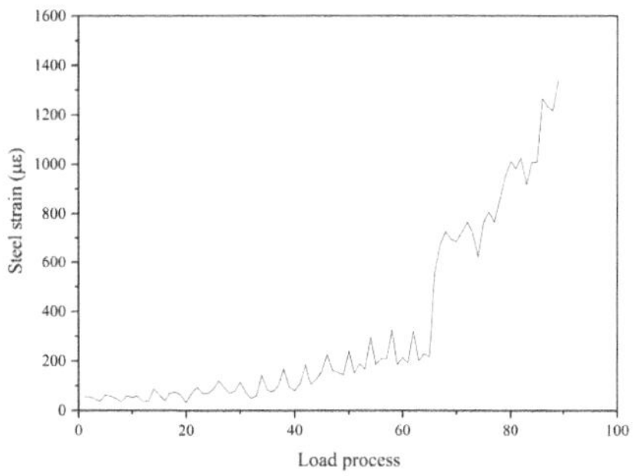

(h) Specimen EHSW3 stirrup strain

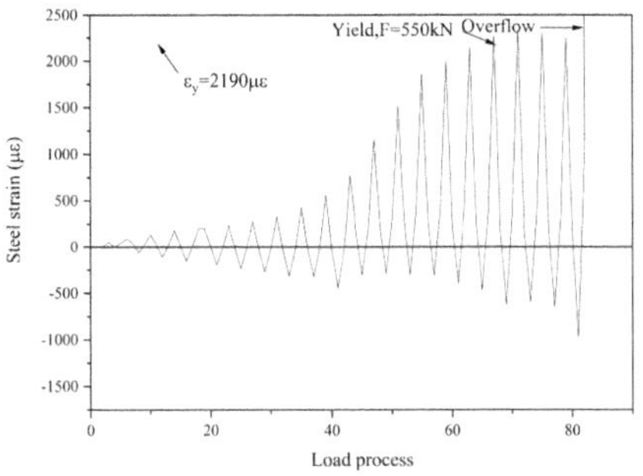

(i) Specimen EHSW4 vertical reinforcement strain

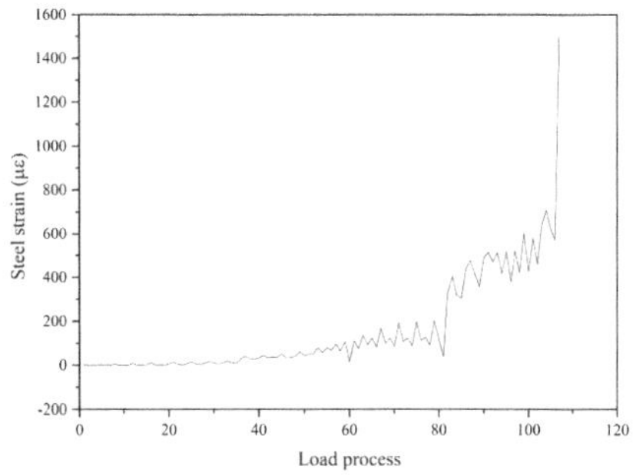

(j) Specimen EHSW4 stirrup strain

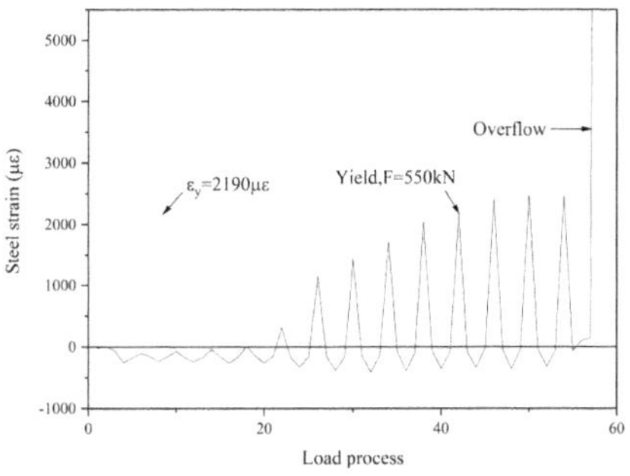

(k) Specimen EHSW5 vertical reinforcement strain

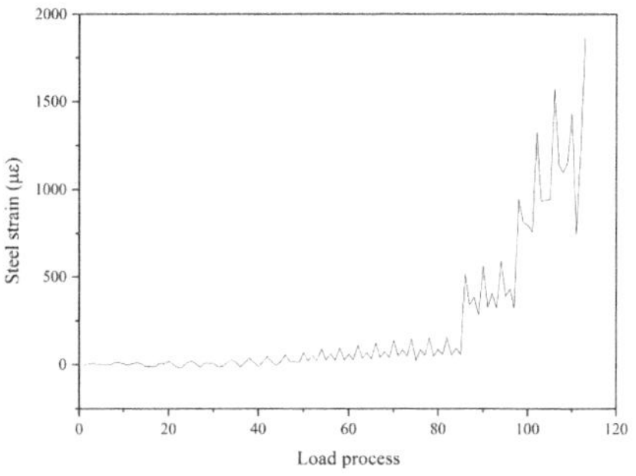

(l) Specimen EHSW5 stirrup strain

Fig. 3.32 (continued)

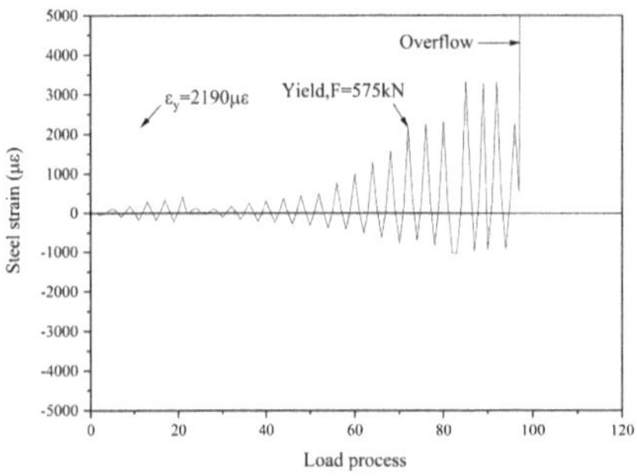

(m) Specimen EHSW6 vertical reinforcement strain

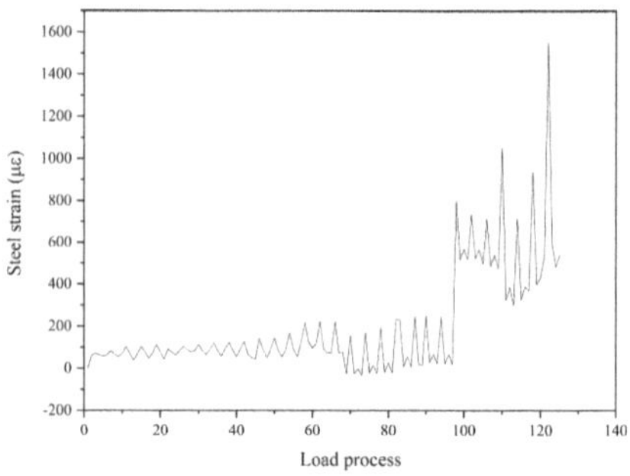

(n) Specimen EHSW6 stirrup strain

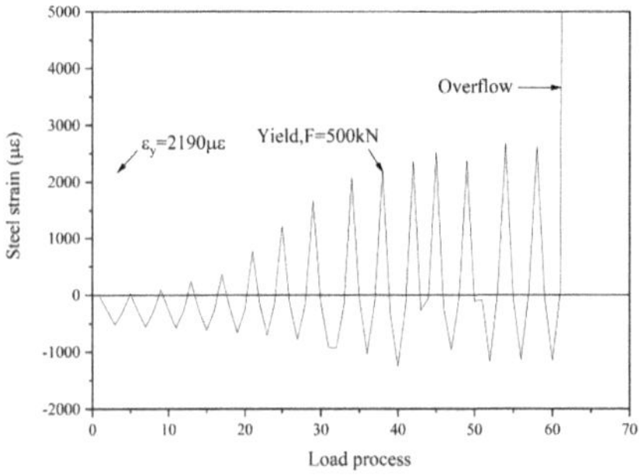

(o) Specimen EHSW7 vertical reinforcement strain

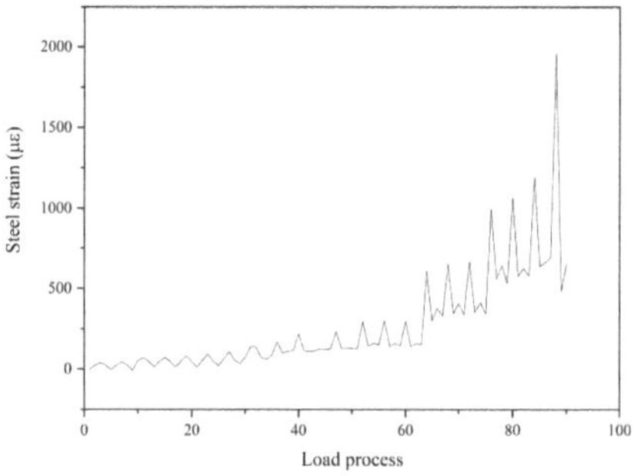

(p) Specimen EHSW7 stirrup strain

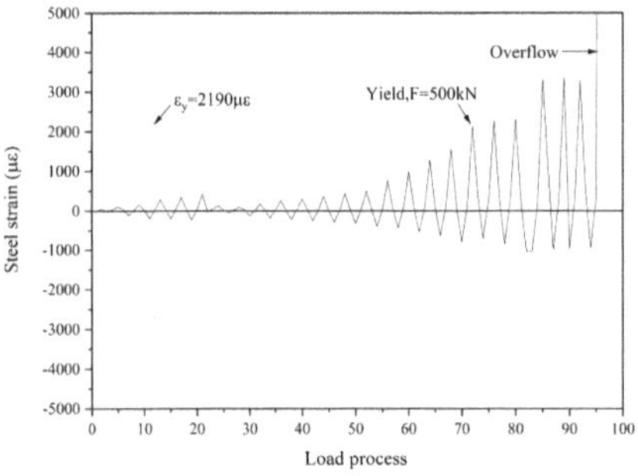

(q) Specimen EHSW8 vertical reinforcement strain

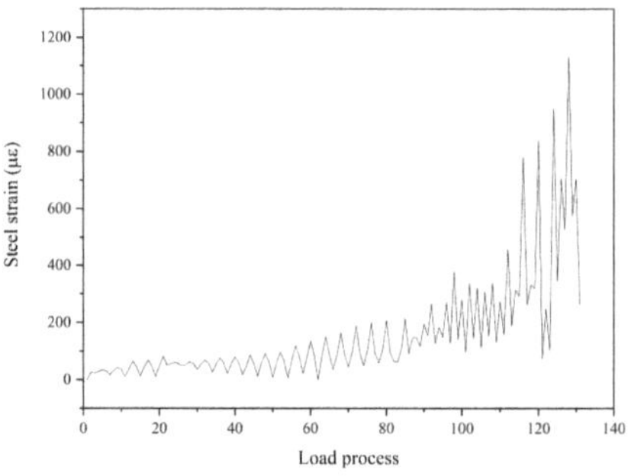

(r) Specimen EHSW8 stirrup strain

Fig. 3.32 (continued)

degree of the specimen concrete is lighter. In contrast, the specimen XJ yields at the first cycle of the fourth stage loading, thus losing constraint on the concrete and causing more serious damage to the concrete.

References

1. National standard of the People's Republic of China. Code for design of concrete structures GB 50010–2010 (2015 Edition). Beijing: China Architecture & Building Press; 2015.
2. Industry standard of the People's Republic of China. Technical specification for concrete structures of tall buildings JGJ 3–2010. Beijing: China Architecture & Building Press; 2010.
3. National standard of the People's Republic of China. Code for seismic design of buildings GB 50011–2010 (2016 Edition). Beijing: China Architecture & Building Press; 2016.
4. National standard of the People's Republic of China. Standard for evaluation of concrete compressive strength GB/T 50107–2010. Beijing: China Architecture & Building Press, 2010.
5. Industry standard of the People's Republic of China. Specification for seismic test of buildings JGJ/T 101–2015. Beijing: China Architecture & Building Press; 2015.
6. Zhu ZF, Guo ZX. Experimental study on seismic performance of new hybrid assembled concrete shear wall with different axial compression ratios [J]. Engineering mechanics, 2019, 33(12): 143–149, 166.
7. Zhu ZF, Guo ZX, Tang L. Seismic performance test of new hybrid assembled concrete shear wall with different unbonded lengths [J]. Engineering mechanics, 2016, 33(8): 52–57.
8. Zhu ZF, Guo ZX, Tang L, Zhu Y. Experiment on seismic performance of new hybrid assembled concrete shear wall considering different pre-tension [J]. Journal of Hunan University (Natural Science Edition), 2015, 42(11): 41–48.
9. Tang JR. Seismic resistance of reinforced concrete frame joints [M]. Nanjing: Southeast University Press, 1989.
10. Zhu ZF, Guo ZX, Tang L. Experimental study and finite element analysis on seismic behavior of new hybrid assembled concrete shear wall [J]. Journal of Civil Engineering, 2018, 51(03): 41–48.
11. Park R. Evaluation of ductility of structures and structural subassemblages from laboratory testing [J]. Bull New Zealand Natl Soc Earthquake Eng., 22(3): 155–166.
12. Industry standard of the People's Republic of China. Standard for seismic design of prestressed concrete structures JGJ/T 140–2019 [S]. Beijing: China Architecture & Building Press, 2019.

Chapter 4 Numerical Analysis of Seismic Performance of Emulative Hybrid Assembled Precast Concrete Shear Wall Based on ABAQUS

4.1 Finite Element Module Method

4.1.1 Analysis Software Selection

Through the nonlinear finite element numerical analysis, it can better simulate the stress state of the structure or component under various actions, study the stress performance of the structure or component, and enrich the experimental research results. In recent years, the finite element analysis method has been widely used in reinforced concrete structures or components. At present, there are many large general software for finite element analysis and calculation, such as ABAQUS, ANSYS, ADINA, MSC. Marc.

Among them, ABAQUS is one of the most advanced large-scale general finite element calculation software in the world, with a wide range of simulation capabilities. ABAQUS has numerous element libraries, material libraries, section libraries, analysis modules, etc., which can be utilized to analyze problems in various fields. It is particularly adept at tackling the nonlinear problems associated with very large and complex reinforced concrete structures, making it widely used in both domestic and foreign research.

Because the EHSW specimen test involves a lot of material nonlinearity (caused by the yield of steel materials and the damage of concrete materials) and boundary nonlinearity (caused by the opening and closing stress state of horizontal joints), this book selects ABAQUS software for finite element analysis.

ABAQUS consists of several functional modules, including the pre-processing module ABAQUS/CAE, the main solver modules ABAQUS/Standard and ABAQUS/Explicit, as well as ABAQUS/Design, ABAQUS/Aqua, ABAQUS/Foundation, etc. In this paper, the main modules used in the finite element analysis process are ABAQUS/CAE and ABAQUS/Standard. Among them, ABAQUS/CAE is the graphical interactive environment of ABAQUS, which can be used to construct specimens conveniently and quickly and display analysis results. ABAQUS/Standard

Z. Zhu et al., *Research on Emulative Hybrid Assembled Precast Concrete Shear Walls*,
https://doi.org/10.1007/978-981-96-9578-2_4

is a general-purpose analysis module that employs an implicit solution method to solve linear and nonlinear problems across a wide range of fields, including static analysis, dynamic analysis, and complex nonlinear coupled physical field analysis. Another solver module, ABAQUS/Explicit, is used for explicit dynamic analysis, which employs explicit methods to solve complex nonlinear dynamic and quasi-static problems, particularly for simulating transient and instantaneous dynamic events [1]. Since the book's test loading method is quasi-static loading, the ABAQUS/Standard analysis module is selected.

4.1.2 Unit Type Selection

ABAQUS has a rich element library. There are over 400 types of elements, divided into 8 categories: continuum elements (solid), shells, membranes, beams, rods, rigid body, connections, and infinites. According to the interpolation order of node displacement, it is divided into linear element, quadratic element and modified quadratic element. The integral algorithm of each element includes a complete integral and a reduced integral [2].

To fully reflect the stress/strain of the specimen, this book selects the continuum element to simulate the specimen. For the concrete component, including the specimen base, wall, and loading beam, C3D8R (8-node hexahedron linear reduction integral element) is used to simulate the three-dimensional stress/strain state of concrete.

Furthermore, to avoid the 'hourglass' numerical problem of linear reduced integration elements, ABAQUS introduces hourglass stiffness in such elements to restrict the expansion of hourglass modes. For reinforcement and prestressed reinforcement, the T3D2 (2-node three-dimensional truss element) is used to simulate the axial stress/strain state of the reinforcement.

4.1.3 Material Constitutive Model

Whether the material constitutive model is correct or not determines the accuracy and credibility of the finite element analysis results, especially for nonlinear analysis of concrete structures. Different descriptions of the material constitutive model can lead to significantly different analysis results, even when using the same model. Therefore, the constitutive model of the material should be chosen appropriately so that the finite element model can accurately reflect the stress and deformation state of the actual structure and obtain correct and available analysis results.

(1) Concrete

Concrete is a widely used material in civil engineering structures. Its essential feature is the inhomogeneity of material composition and the presence of initial micro-cracks. The uniaxial compression stress–strain relationship of concrete reveals residual deformation upon unloading, which does not adhere to the elastic relationship. If the elastoplastic constitutive relation is applied, it is difficult to accurately define the yield condition. In addition, after the concrete reaches the peak stress, the stress–strain relationship curve appears the descending section, and there is strain softening phenomenon. All the above factors have brought difficulties in establishing a concrete constitutive model. Numerous scholars have conducted extensive experimental and theoretical research and have proposed a variety of concrete constitutive models. These can be roughly divided into the linear elastic constitutive model, the plastic theory constitutive model, the other mechanical theory constitutive model, and the nonlinear elastic constitutive model [3].

ABAQUS provides three basic constitutive models for concrete materials: the Brittle Cracking Model, the Smeared Cracking Model, and the Damaged Plasticity Model [2]. The three models belong to other constitutive models of mechanical theory. Among them, the brittle cracking model only considers the nonlinear behavior of concrete under tension, which is suitable for simulating the constitutive relationship of concrete materials in plain or lightly reinforced concrete structural members. However, it is not applicable for simulating concrete materials in normal reinforced concrete structures or composite structures. The dispersion cracking model is primarily suitable for the analysis of low confining pressure and monotonic loading of various reinforced concrete structures and can also be used for plain concrete. The damage plasticity model is suitable for simulating the constitutive relation of concrete materials under cyclic loading and can consider the damage, crack development, and stiffness recovery of materials under cyclic loading. Therefore, based on the characteristics of this model test analysis, this book selects the damage plasticity model for analysis.

The damage plasticity model provided by ABAQUS is based on the model proposed by Lubliner et al., which introduces a damage index into the concrete constitutive model to reduce the elastic stiffness matrix of concrete. This enables the simulation of the characteristic that the unloading stiffness of concrete decreases with increasing damage, especially for nonlinear analysis of concrete [4].

According to the uniaxial tensile/compressive stress–strain curve of concrete considering damage factor given in Fig. 4.1, the mathematical description of stress, strain and damage factor of concrete is shown in Eqs. 4.1–4.3.

$$\sigma = (1 - d)\overline{\sigma} \tag{4.1}$$

$$\overline{\sigma} = D_0^{el}\left(\varepsilon - \widetilde{\varepsilon}^{pl}\right) \tag{4.2}$$

$$\sigma = (1 - d)D_0^{el}\left(\varepsilon - \widetilde{\varepsilon}^{pl}\right) \tag{4.3}$$

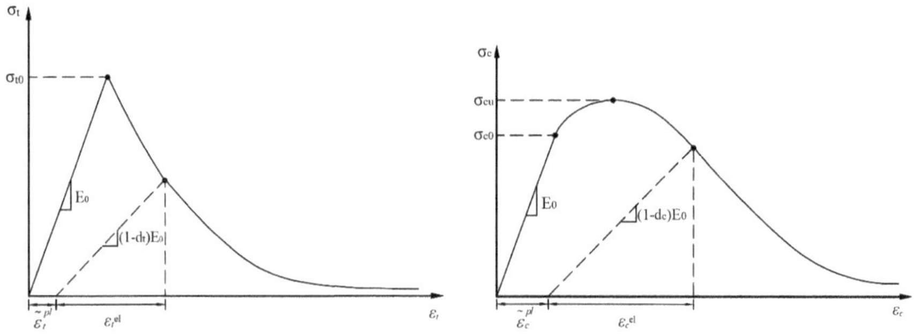

(a) uniaxial tension (b) uniaxial compressive

Fig. 4.1 Uniaxial tensile/compressive stress–strain curves of concrete considering damage factor [2]

d Damage factor, varying from 0 (no damage) to 1 (complete failure), represents the stiffness degradation of concrete due to damage under tension and compression, respectively.

$\overline{\sigma}$ Effective stress,

D_0^{el} The initial (lossless) stiffness of the material can also be written as E_0,

$\widetilde{\varepsilon}^{pl}$ Plastic strain.

The damage factor in the damage plasticity model is a key issue, and as can be seen from Formulas 4.1–4.3, the calculation of the damage factor is closely tied to the elastic–plastic stress–strain curve of concrete. For the calculation of damage factor, References [5] and [6] gave the method of inverse calculation according to Sidiroff energy equivalence principle or stress–strain relationship. In this book, the constitutive curves suggested by Appendix C.2.3–C.2.4 of 'Code for design of concrete structures' (GB50010-2010) [7] are selected, and the damage factors of the concrete constitutive model are calculated according to Eqs. 4.4–4.9 According to Reference [8], the damage factors under tensile and compressive conditions are calculated separately because of the significant differences between the tensile and compressive properties of concrete.

$$d_c = 1 - \frac{\sigma_c E_0^{-1}}{\widetilde{\varepsilon}_c^{pl}(1/b_c - 1) + \sigma_c E_0^{-1}} \tag{4.4}$$

$$b_c = \widetilde{\varepsilon}_c^{pl}/\varepsilon_c^{in} \tag{4.5}$$

$$\varepsilon_c^{in} = \varepsilon_c - \sigma_c/E_0 \tag{4.6}$$

$$d_t = 1 - \frac{\sigma_t E_0^{-1}}{\widetilde{\varepsilon}_t^{pl}(1/b_t - 1) + \sigma_t E_0^{-1}} \tag{4.7}$$

$$b_t = \tilde{\varepsilon}_t^{pl} / \varepsilon_t^{in} \tag{4.8}$$

$$\varepsilon_t^{in} = \varepsilon_t - \sigma_t / E_0 \tag{4.9}$$

$\varepsilon_c^{in}, \varepsilon_t^{in}$ Unloading residual or plastic strain without considering damage,
b_c, b_t All tests, by cyclic load-unloading–reloading stress path calibration, followed literature recommendations [9], with amplitudes of 0.7 and 0.1.

In addition, plasticity must be defined by true stress and true strain in ABAQUS. However, most experimental data are often provided in terms of nominal stress and nominal strain values. Therefore, it is necessary to transform the nominal stress (change) of plastic materials into real stress (change). The specific process is shown in Formulas 4.10 and 4.11.

$$\sigma = \sigma_{nom}(1 + \varepsilon_{nom}) \tag{4.10}$$

$$\varepsilon = \ln(1 + \varepsilon_{nom}) \tag{4.11}$$

ε_{nom} Nominal strain of concrete, i.e. strain of measured or theoretical stress–strain curve of concrete,
σ_{nom} The nominal stress of concrete is the stress of measured or theoretical stress–strain curve of concrete.

Therefore, based on the characteristic values of concrete material given in Table 3. 2, the important curves in the plastic damage constitutive model of concrete in ABAQUS can be obtained through the calculations of Formulae 4.1–4.11, i.e., the inelastic stress–strain curve and the damage factor-inelastic strain curve under tension and compression. The relevant curves are detailed in Fig. 4.2.

For uniaxial cyclic loading (tension–compression-tension), a weight factor is (tension, compression, see Fig. 4.3) introduced into the ABAQUS plastic damage constitutive model, which is related to the material properties. It is used to describe the degree of stiffness recovery under reverse load. The value ranges from 0 to 1, representing the degree of stiffness recovery when the material changes from compression to tension. A value of 1 indicates complete recovery of stiffness, while a value of 0 means that the stiffness is not recovered. A value between 0 and 1, but greater than 0, indicates partial recovery of stiffness.

Furthermore, the plastic damage constitutive model of concrete in ABAQUS requires the definition of a series of plastic parameters to determine the material's yield criterion and flow rule. The plastic damage constitutive model adopts a Lubliner yield surface and a hyperbolic DP flow potential energy surface. The shape of the hyperbolic DP flow potential energy surface on the meridian plane is defined by the Dilation Angle and the Eccentricity of the plastic potential equation. Among them, the expansion angle has a significant impact on the behavior of concrete, and its value should be chosen carefully; the flow angle can generally be set to the default

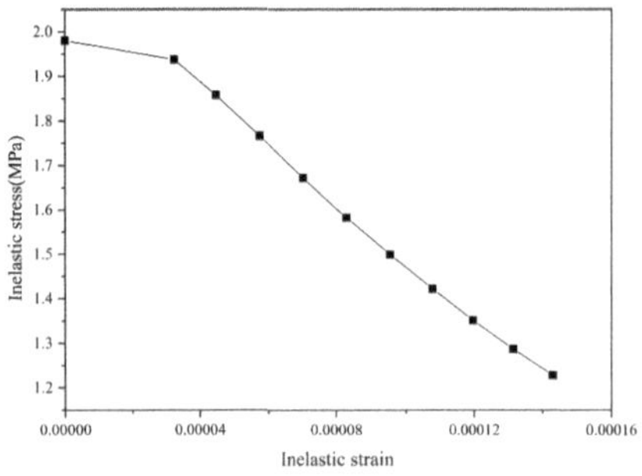

(a) Tensile inelastic stress-strain curve

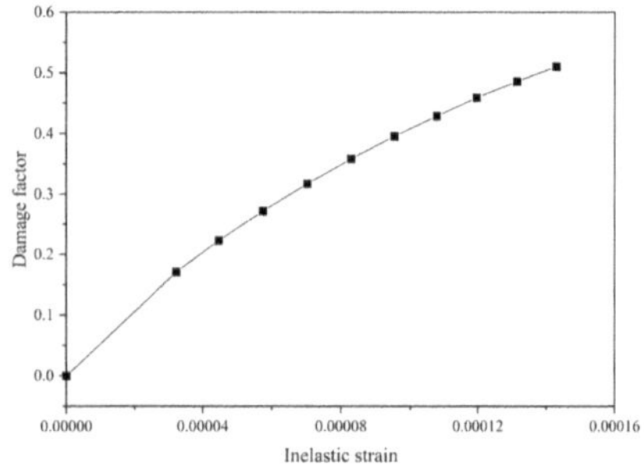

(b) Tensile damage - inelastic strain curve

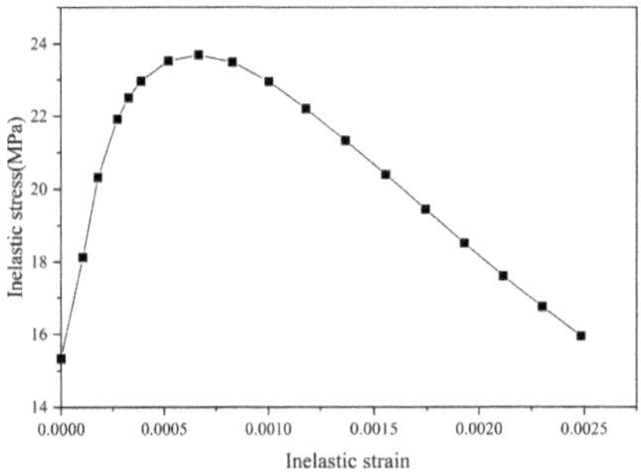

(c) Compression inelastic stress-strain curve

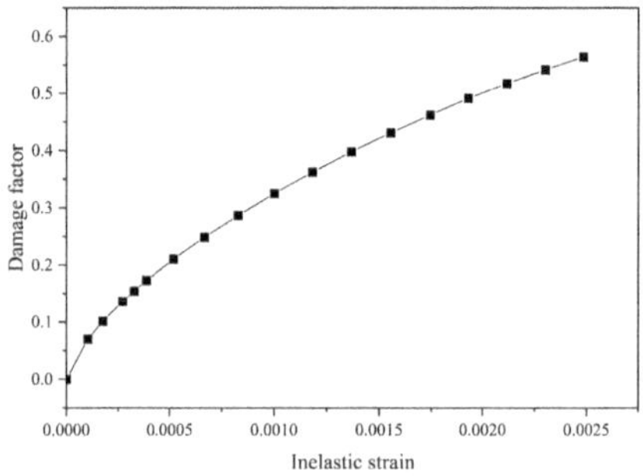

(d) Compression damage - inelastic strain curve

Fig. 4.2 Constitutive model curve of concrete in ABAQUS

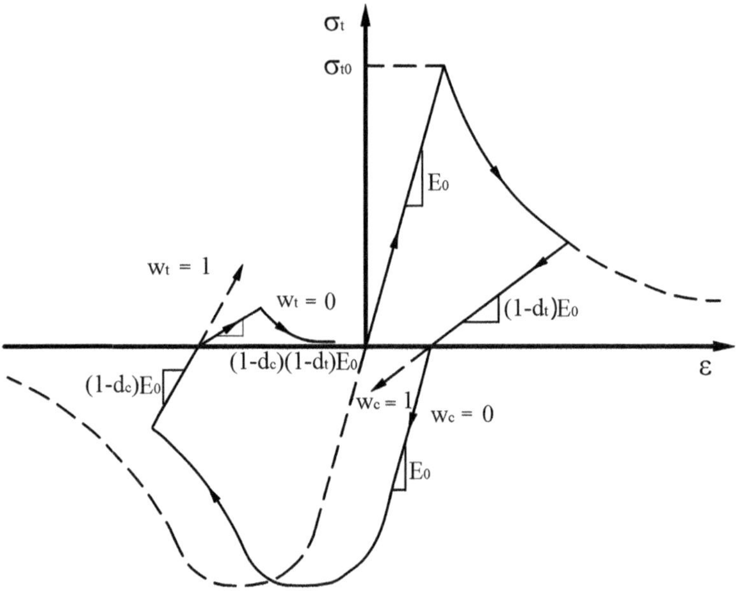

Fig. 4.3 ABAQUS constitutive model curve under uniaxial cyclic loading [2]

Table 4.1 Plastic parameters of ABAQUS constitutive model

Parameter	Angle of dilatancy	Eccentricity	f_{b0}/f_{c0}	K_c	λ
Values	55	0.1	1.16	0.6667	0.0005

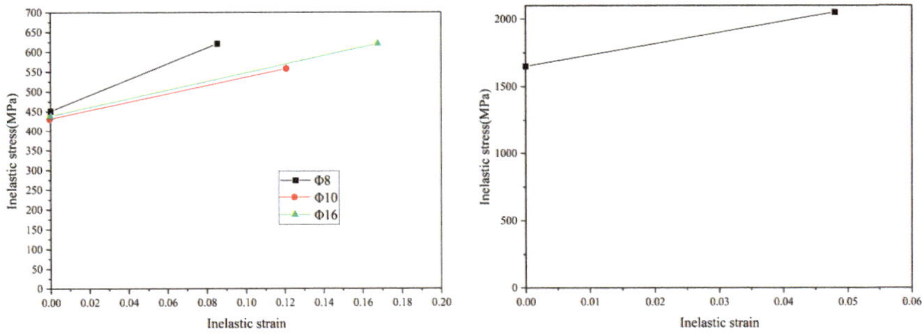

(a) ordinary rebar (b) prestressed reinforcement

Fig. 4.4 The constitutive model curve of steel bar in ABAQUS

value of 0.1. The shape of the yield surface on the deviatoric plane and the plane stress plane is defined by the ratio of the equivalent biaxial compressive stress to the uniaxial compressive stress (f_{b0}/f_{c0}) and the ratio of the tensile meridian to the second stress invariant on the compressive meridian (K_c).The default values are 1.16 and 2/3, respectively. The viscosity coefficient is determined by λ, and the default value is generally 0.0005 [6]. Through multiple trials, considering both solution convergence and result consistency, the relevant parameters of the model in this book are presented in Table 4.1.

(2) steel reinforcement

Reinforcement materials are typically modeled using a simple bilinear elastoplastic model, and strain hardening can be accounted for with a constant slope after yielding. In this book, bilinear simplification is carried out based on the measured curve of steel bar material test, and the conversion of real stress (variable) is performed using Formulas 4.10 and 4.11. Finally, the constitutive model curve of steel bar material (including both ordinary steel bar and prestressed steel bar) is presented in Fig. 4.4.

4.1.4 Related Interface Simulation

(1) Interface between ordinary steel and concrete Related interface simulation

For different treatment methods between steel and concrete, the general finite element model of reinforced concrete can be divided into three types: distributed model, separated model, and embedded model (also known as the combined model). The distributed reinforcement model disperses the reinforcement within the concrete element, and the contribution of the reinforcement to the structure is considered by adjusting the material properties of the element. The separate model simulates the bond-slip relationship by discretizing the steel and concrete elements and adding a joint element at the interface. The separate model can directly reflect the interaction between steel bar and concrete and obtain the crack distribution more accurately. However, it requires separate displacement degrees of freedom for the steel bar and concrete, resulting in a complex grid and large calculation volume. For complex reinforced concrete structures, the calculation efficiency of the separate model is low. The embedded model is based on the non-slip assumption, and the element stiffness is obtained by the superposition of the stiffness components of steel bar and concrete, ignoring the influence of interface mechanical action on structural stiffness. Embedded models are divided into two types: one is a layered combination, and the other is a reinforced concrete composite unit.

For the finite element analysis of the test model, an embedded analysis model is utilized to differentiate between steel and concrete elements, with the steel element embedded within the concrete element. To ensure calculation accuracy and efficiency, the bond slip effect between steel and concrete is neglected. In ABAQUS, the reinforcement element is embedded into the concrete element for simulation by establishing an embedded constraint relationship between the reinforcement and concrete elements. For the unbonded segment, it forms an independent unit with the bonded segment and does not establish any constraint relationship with the concrete to simulate its unbonded effect with the concrete.

(2) Interface between unbonded prestressing tendons and concrete

Considering that there is no bond between the unbonded prestressed tendon and the model concrete, there is no connection between the unbonded prestressed tendon and the concrete in the model, but the X and Y directions between the unbonded prestressed tendon unit node and the nearby concrete unit node are constrained by the coupling equation. Fixed, and only Z-direction deformation occurs, that is, the axial elongation or shortening of the prestressed tendon is allowed, and the prestressed tendon pipeline cannot be deviated. For the anchorage points at both ends of the prestressed tendon, it is connected by the coupling equation (Coupling) between the end node of the prestressed tendon and the top surface of the loading beam and the bottom surface of the base.

(3) Horizontal seam interface

It can be found from the previous test that there is no bond anchorage failure in the grouted steel bar. Therefore, it can be considered that the existence of EHSW joint interface is an important reason for the difference in seismic performance between EHSW specimen and specimen XJ. In the finite element model, it must be correctly simulated to truly reflect the mechanical properties of the model seam, so as to ensure the reliability of the overall model.

In ABAQUS, the simulation of the seam interface is treated as contact and defined in ABAQUS/Standard. The contact simulation is either based on the surface or based on the contact element. The book adopts a surface-based surface-to-surface contact simulation method, eliminating the need for introducing additional contact elements and reducing calculation consumption while ensuring calculation accuracy. The interaction between the contact surfaces includes two parts, one is the normal action between the contact surfaces, the normal action includes the contact and separation of the contact surfaces and the normal stress; the other part is the tangential action between the contact surfaces, which includes the relative motion (sliding) between the contact surfaces and the possible frictional shear stress. Each contact interaction can represent a contact characteristic, which defines a model of interaction between contact surfaces. In the ABAQUS model of this book, the default hard contact is used for the normal behavior of the contact surface. That is, when the gap is zero (the contact surfaces are in contact), the contact constraint is applied. When the contact pressure between the surfaces becomes zero or negative, the surfaces separate, and the applied constraint is removed. For tangential behavior of the contact surface, the classical Coulomb friction model is used to simulate the friction coefficient. The friction coefficient between rough concrete surfaces is 0.6, and the maximum friction shear stress is limited to the tensile strength of concrete material (2.5 MPa) [9].

1) Interface of Corrugated Pipe, Grouting Material and Concrete

The failure phenomenon of the interface between the bellows, the grouting material and the concrete does not occur during the test. At the same time, considering the existence of higher strength grouting materials, the strength of compressed concrete will be improved. Additionally, the bellows will further constrain the slightly brittle grouting material and improve its ductility. To facilitate modeling, the finite element model does not directly incorporate a bellows model or grouting material; instead, a concrete material unit is used.

4.1.5 Element Partition

The size and shape of the element mesh have a direct impact on computational accuracy and efficiency. The smaller the mesh size, the higher the computational accuracy, but the higher the computational cost, and vice versa; the grid shape is particularly important for calculation accuracy. A grid with high shape quality not only has high accuracy, but also enables fast calculations.

For the mesh size, combined with the characteristics of the concrete model analysis, the mesh size is controlled to 50 mm. It should be noted that due to the lack of attention to stress/strain of the base part, the unit size of the base part is appropriately increased and adjusted according to the convergence of the calculation results.

For the grid shape, the concrete component includes the concrete base, wall, and loading beam, all using hexahedral elements; the steel bar component includes both ordinary and prestressed steel bars, represented using line elements.

For meshing technology, to adapt to the determined mesh size and shape, the structured meshing technology of ABAQUS is adopted through topological operations such as segmentation and merging of geometric models.

4.1.6 Boundary Conditions and Loading

The simulation of boundary conditions directly impacts the correctness of the calculation results for the overall finite element model. Therefore, the finite element model must be constrained appropriately based on the real boundary conditions of the test model. The test models in this book are anchored to the laboratory ground using pre-tensioned finished rolled rebar to simulate the constraint conditions of the fixed support at the bottom of the shear wall. Therefore, the bottom surface of the base is completely fixed in the finite element model of this book to approximate the real test boundary conditions.

For the loading method, for the axial pressure, the loading beam top surface load is applied to simulate; in the finite element model, the requirements of computational convergence and computational cost are taken into account, and the existing correlation analysis shows that the load–displacement curve under monotonic horizontal load is in good agreement with the envelope of the hysteresis curve under low cyclic loading. Therefore, the lateral resistance of the model is analyzed by applying a monotonic horizontal load. At the same time, to simulate the real loading tooling conditions, the horizontal load is applied to the reference point in the model. The reference point is located at the center of the loaded beam and connected to the end face of the loaded beam through the coupling constraint condition to avoid the influence of stress concentration on the loading and calculation results. In addition, the loading is still controlled by displacement, and the ultimate displacement is the same as the test termination displacement, which is set to 93 mm.

4.1.7 Finite Element Model

Through the above process, the finite element model corresponding to the test model can be established respectively. Among them, in view of the test did not occur in the loading beam and the base damage phenomenon, while in order to save computing costs, in addition to the shear wall, for the concrete material constitutive model, a linear elastic constitutive model was used. In addition, for the convenience of modeling, unbonded prestressed tendons are simulated by a separate line entity, and the change of its number is simulated by the change of the cross-sectional area of the line entity.

The finite element models of XJ and EHSW specimens are shown in Fig. 4.5.

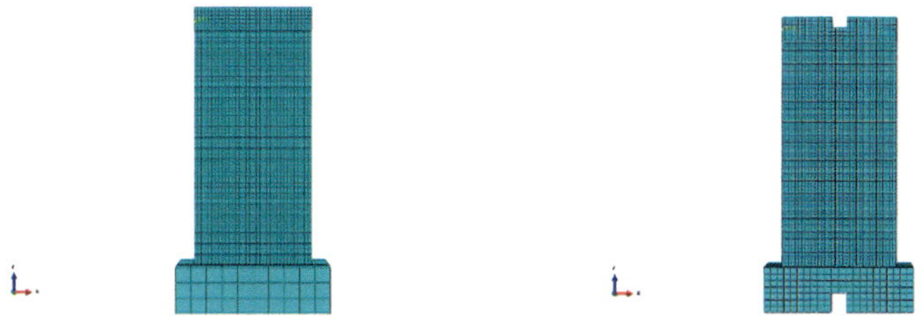

(a) Specimen XJ (b) specimen EHSW

Fig. 4.5 ABAQUS finite element model

4.2 Finite Element Model

To ensure the accuracy of the finite element model and the reproducibility of the test results, the finite element analysis results are compared with the test results, including the load–displacement curve, prestressed tendon stress, and model failure mode.

4.2.1 Comparison of Load–Displacement Curves

Since the finite element model only applies monotonic horizontal load, only the unidirectional monotonic load–displacement curve can be obtained. The analysis result curve is compared with the experimental hysteresis curve to determine if the calculated load–displacement curve forms a good envelope of the measured hysteresis curve.

The measured hysteretic curves of each test model and the monotonic load–displacement curves from finite element analysis are presented in Fig. 4.6. The monotonic load–displacement curve obtained by finite element analysis can better envelop the measured hysteresis curve of the test, especially in the displacement loading stage, as can be seen from the diagram. The analysis results are in good agreement with the periphery of the test hysteresis curve. At the same time, the prediction of the peak load by finite element analysis is close to the measured results, with the maximum error controlled within 7%.

At the same time, it can be found that the finite element analysis results show higher stiffness in the elastic stage, resulting in a more obvious deviation between the analysis curve and the measured curve. The analysis shows that this phenomenon is mainly caused by the influence of the slight gap between the specimen and the ground and the loading equipment, which exists in the actual test but cannot be considered by ABAQUS. With increasing loading grade, specimen deformation gradually increases, and gap influence gradually decreases, such that analysis curve and test curve show good approximation effect in displacement loading stage [9].

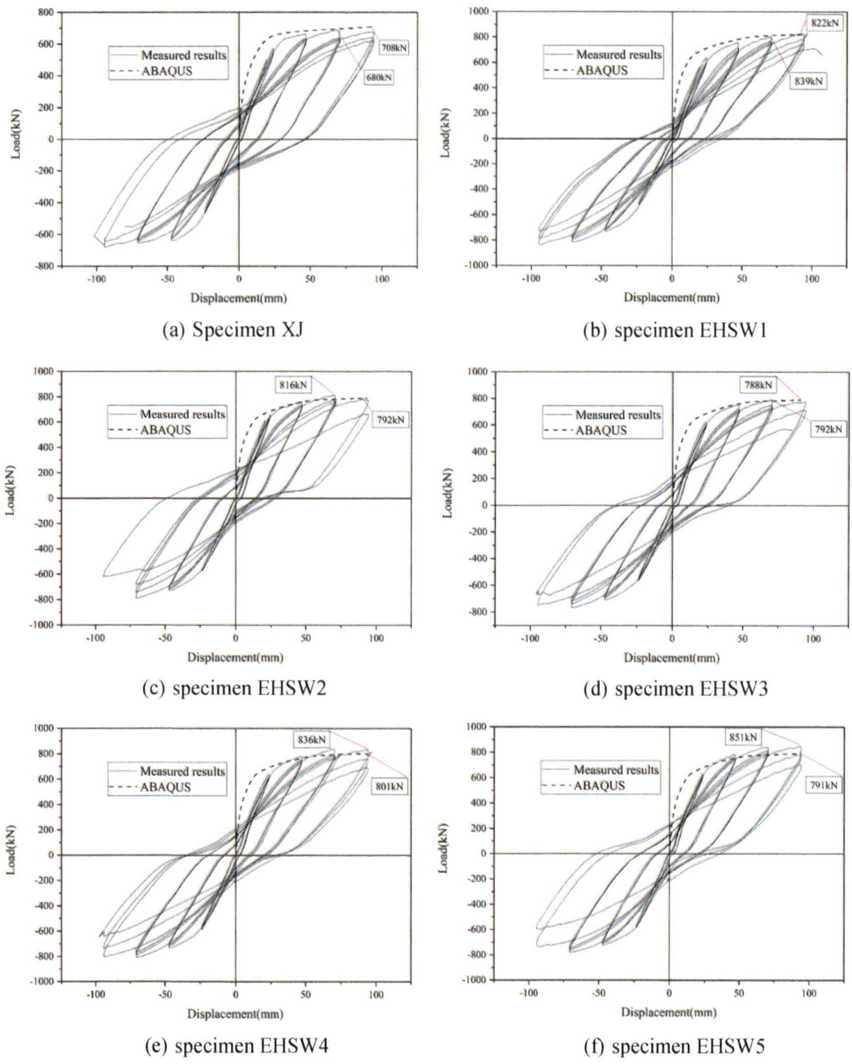

(a) Specimen XJ

(b) specimen EHSW1

(c) specimen EHSW2

(d) specimen EHSW3

(e) specimen EHSW4

(f) specimen EHSW5

Fig. 4.6 Comparison of load–displacement curves

4.2.2 Stress Comparison of Prestressed Tendon

Compared to specimen XJ, the influence of the stress state of the prestressed tendons of the EHSW specimen on the overall lateral resistance of the model is particularly important. Therefore, the simulation degree of prestressed tendon stress state will directly determine whether the finite element model is correct or not. Table 4.2 presents a comparison between the calculated values and the measured values of the stress and peak stress of the prestressed tendons after axial compression. As can be

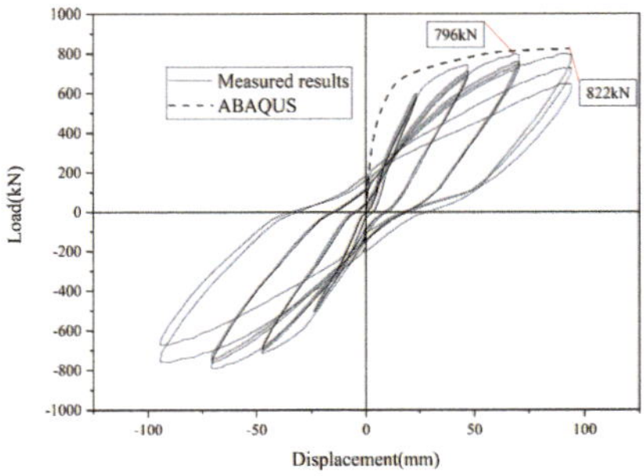

(g) specimen EHSW6

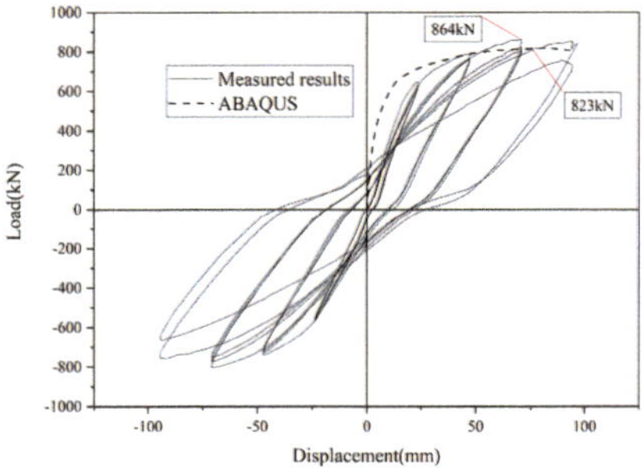

(h) specimen EHSW7

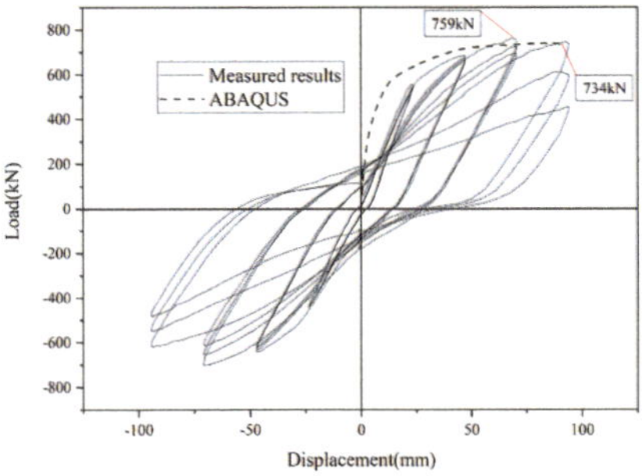

(i) specimen EHSW8

Fig. 4.6 (continued)

seen from the table, the calculated values are generally close to the measured values, and the maximum absolute error is within 5%.

4.2.3 *Comparison of Model Failure Modes*

Through testing, it can be found that the final failure modes of specimens XJ and EHSW are basically the same, both of which are bending shear failure at the root of the shear wall. The plastic failure of the concrete is concentrated at the end of the shear wall and is limited to a certain height range from the joint or the top surface of the base. For the finite element model under monotonic horizontal load, the corresponding plastic strain of concrete is concentrated on the compression side of the wall and near the root of the wall. As shown in Fig. 4.7, consistent with experimental results, finite element analysis results also yield a considerable model failure mode.

Table 4.2 Stress comparison of prestressed tendon

Test piece	Initial prestress/ MPa	Prestress after axial compression/ MPa			Peak stress/MPa		
		Measured value	Calculated value	Error (%) [1]	Measured value	Calculated value	Error (%)[1]
EHSW1	880.3	864.2	866.4	0.25	1158.9	1149.5	– 0.81
EHSW2	758.5	741.8	746.3	0.61	1082.1	1038.5	– 4.03
EHSW3	1266.1	1248.2	1253.5	0.42	1416.1	1444.9	2.03
EHSW4	676.8	658.9	664.9	0.91	948.2	933.3	– 1.57
EHSW5	530.3	514.2	518.6	0.86	808.9	812.3	0.42
EHSW6	827.6	811.5	813.7	0.27	1127.6	1085.2	– 3.76
EHSW7	893.7	877.6	879.8	0.25	1177.6	1149.8	– 2.36
EHSW8	885.2	877.1	875.6	– 0.17	1167.4	1199.0	2.71

Note 1.Error = (calculated value-measured value)/measured value × 100%

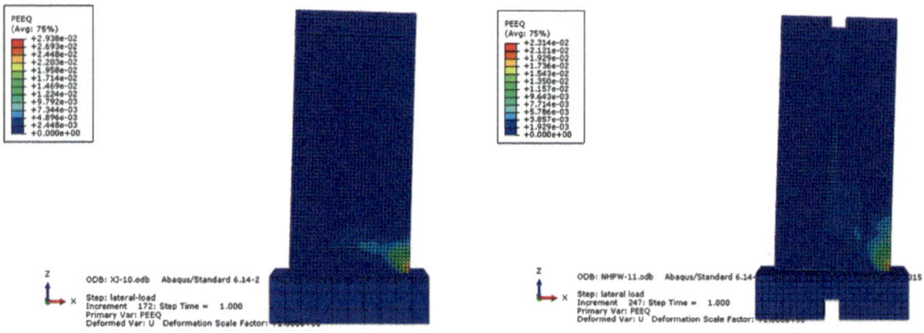

(a) specimen XJ (b) specimen EHSW

Fig. 4.7 Comparison of model failure modes

4.3 Finite Element Parameter Analysis

Through mutual verification of finite element analysis results and test results, reproducibility of the finite element model to the test is demonstrated, and accuracy of the finite element analysis results is ensured. On this basis, based on the verified finite element model, the basic simulation scheme of the model is kept unchanged, and the relevant parameter analysis is further carried out. The parameter types still include unbonded prestressed tendon pretension, unbonded prestressed tendon area, unbonded length of slurry anchor steel bar and axial compression ratio. Four types of parameters are considered. The values of parameters are taken into account that the test does not take into account or that the actual specimen test cannot be realized. By comparing the monotonic load–displacement curves of each parameter change model, the influence of each key parameter on the bearing capacity, stiffness and other lateral resistance of EHSW specimens is further mastered.

4.3.1 Analysis of Pretension Parameters of Unbonded Prestressed Tendons

For the pretension parameters of unbonded prestressed tendons, three initial prestressing conditions of 530.3 MPa ($0.29f_{ptk}$), 676.8 MPa ($0.36f_{ptk}$) and 880.3 MPa ($0.47f_{ptk}$) were considered by the test model groups EHSW5, EHSW4 and EHSW1, respectively. In order to further explore the influence of this parameter on the lateral resistance of EHSW specimens, a series of different initial prestress models is added. The specific information of each model is detailed in Table 4.3. The comparison of analysis results for each parameter change model is shown in Fig. 4.8.

It can be seen from Fig. 4.8 that the size of the unbonded prestressing tendon pretension has a significant effect on the monotonic load–displacement curve of the model. With increasing prestressing tendon pretension, the curve shifts upward, and the stiffness and bearing capacity of the model are significantly improved. This phenomenon has not been fully reflected in the model test, especially the impact on strength. Due to the testing error, the discrimination of the test results is low, failing to fully reflect the influence of pre-tension on the unbonded prestressed tendon on the strength of the model. Finite element analysis results indicate that increasing the pretension of prestressed tendons is an effective means to improve the stiffness and bearing capacity of the EHSW model.

Table 4.3 Pretension parameter analysis model of unbonded prestressed tendons

Model	Initial prestress/MPa	Number of prestressing tendons	Tendons area/mm^2	Pretension/kN	Remark
P-1	0	4	560	0	–
P-2	279 ($0.15f_{ptk}$)	4	560	156.2	–
P-3	530.3 ($0.29f_{ptk}$)	4	560	297.0	Test model EHSW5
P-4	676.8 ($0.36f_{ptk}$)	4	560	379.0	Test model EHSW4
P-5	880.3 ($0.47f_{ptk}$)	4	560	493.0	Test model EHSW1
P-6	1116 ($0.60f_{ptk}$)	4	560	625.0	–
P-7	1395 ($0.75f_{ptk}$)	4	560	781.2	–

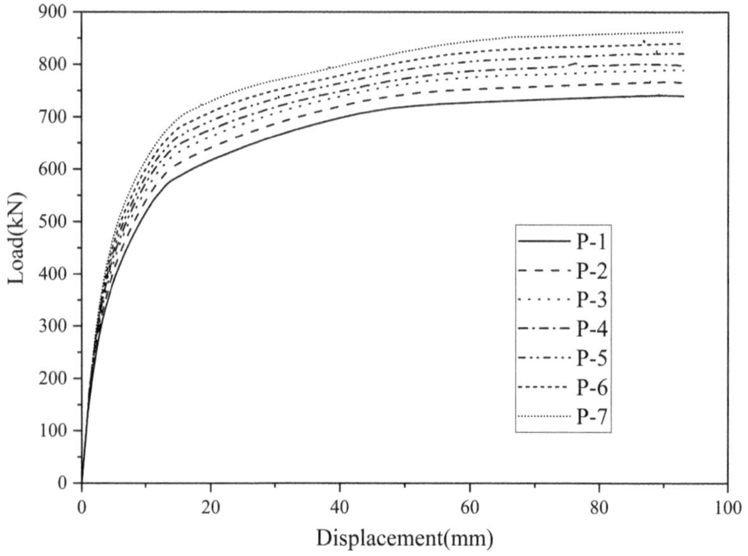

Fig. 4.8 Analysis results of pretension parameters of unbonded prestressed tendons

4.3.2 Unbonded Prestressed Tendon Area Parameter Analysis

Three types of area changes for 2, 3, and 4 unbonded prestressed tendons were considered in the test. For the test model, it is basically enough to consider the change of parameters (the larger the area, the larger the required channel diameter, the higher the thickness of the model wall, and the possible wall size is not enough; the smaller the area, the greater the initial prestress of the prestressed tendon under the premise of the same pretension, which may exceed the nominal yield strength of the prestressed tendon). However, it can be found through experiments that due to the process error and equipment error during the prestressing of the prestressed tendon, the pre-conceived changes in the area of the prestressed tendon and the pretension remain unchanged cannot be accurately realized (see Table 3.17). Therefore, the influence of the area of unbonded prestressed tendons summarized by the test results on the seismic performance of the model is bound to be affected by different pretensions. In the finite element model, by setting the initial stress field, the pretension can be well consistent, allowing for a more intuitive and reliable discussion of the model's performance with respect to this parameter.

The finite element models considering different prestressed tendon areas are listed in Table 4.4, and the comparison of the analysis results of each parameter variation model is shown in Fig. 4.9.

From Fig. 4.9, it can be found that, consistent with the law shown in the test results, under the premise of keeping the pretension of unbonded prestressed tendons unchanged, the influence of the change of prestressed tendon area on the monotonic load–displacement curve is not obvious. The curves basically coincide at the initial

Table 4.4 Area parameter analysis model of unbonded prestressed tendons

Model	Number of prestressing tendons	Tendons area/mm^2	Initial prestress/MPa	Pretension/kN
A-1	2	280	1116	312.5
A-2	3	420	744	312.5
A-3	4	560	558	312.5

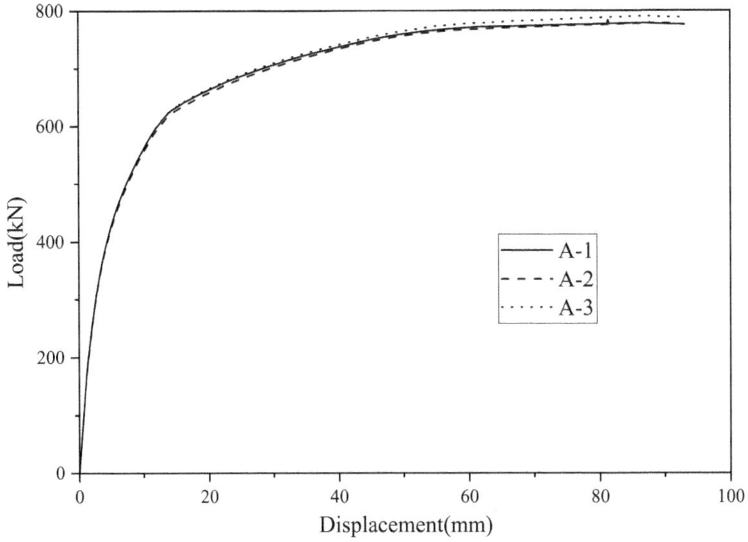

Fig. 4.9 Analysis results of unbonded prestressed tendon area parameters

stage of loading, and the curves are separated from each other until the later stage of loading. With the increase of prestressed tendon area, the bearing capacity is improved to a certain extent, but it is still not obvious. According to the analysis, although the increase of the area of unbonded prestressed tendons can improve the reinforcement ratio of the model, because the prestressed tendons are located in the middle of the model, the bending arm is small, and the contribution to the bending capacity of the model is relatively small. Therefore, while an increase in the area of prestressed tendons can improve the bearing capacity and stiffness of the model in the later stage, the amplitude is small.

4.3.3 *Parameter Analysis of Unbonded Length of Grouted Steel Bar*

In the test, the change of unbonded length of 150, 200 and 250 mm three kinds of slurry anchor steel bars is considered. To further understand the influence of unbonded length in a larger range on the lateral resistance of the model, 0 mm (that is, all slurry anchor steel bars are bonded) and 500 mm are added here. The finite element model numbers are sequentially labeled as U-1 to U-5 according to the increasing length of the unbonded segment. The comparison of the analysis results of each parameter change model is shown in Fig. 4.10.

It can be seen from Fig. 4.10 that the influence of the unbonded length of the grout-anchored steel bar on the strength and stiffness of the model does not reflect obvious regularity, but it can still be seen that the increase of the unbonded length will gradually shift the inflection point of the curve to the positive direction of the displacement axis, which makes the yield of the model lag behind, resulting in an increase in the yield load, but the peak load does not show the same regularity as the test. The analysis indicates that due to the limitations of the finite element model, the spalling and crushing of concrete on the compression side and the bond slip between the steel bar and concrete cannot be fully considered, leading to relatively inaccurate simulation results of the stress state of unbonded grouted anchor steel bars on the tension side. This, in turn, affects the variation law of the model's strength to a certain extent. Especially under large deformation conditions, the error will be greater when the plasticity of the steel bar is fully developed.

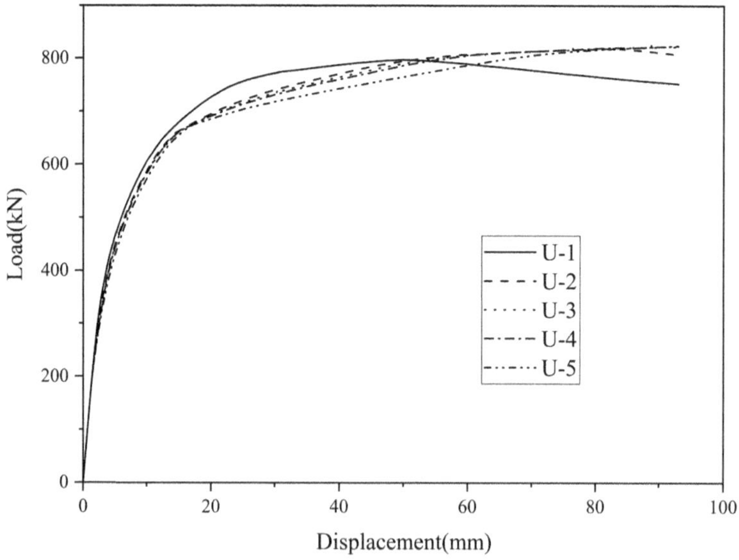

Fig. 4.10 Analysis results of unbonded length parameters of grouted anchor steel bar

4.3.4 Parameter Analysis of Axial Compression Ratio

Due to the limitations of the test device and laboratory conditions, the axial compression ratio in the model test cannot be too high. After the conversion of axial compression ratio, the design axial compression ratio that can be considered in the test of this book is 0.12 and 0.24, which is still less considered for shear wall members in general structures. Therefore, an additional model of axial compression ratio parameter change is incorporated here. The axial compression ratio information for the model is presented in Table 4.5. The comparison of analysis results for each parameter change model is shown in Fig. 4.11.

Table 4.5 Parameter analysis model of axial compression ratio

Model	Test axial compression ratio	Design axial compression ratio	Axle pressure/kN	Remark
AL-1	0	0	0	–
AL-2	0.055	0.12	473.15	Test model EHSW8
AL-3	0.11	0.24	946.3	Test model EHSW1
AL-4	0.2	0.44	1720.5	–
AL-5	0.3	0.65	2580.8	Design axial compression ratio exceeds axial compression ratio limit 0.5

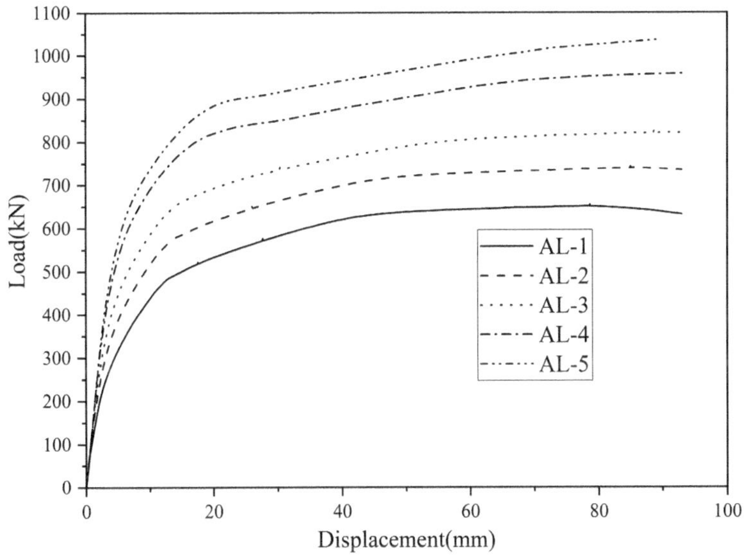

Fig. 4.11 Parameter analysis results of axial compression ratio

From Fig. 4.11, it can be seen that, similar to the initial prestress parameters of unbonded prestressed tendons, the increase of axial compression ratio will significantly improve the stiffness and bearing capacity of the EHSW model. However, it can also be seen from the inflection point of the curve that the increase of axial compression ratio will lead to the delay of curve bending, that is, the yield of the model will be significantly delayed, which may be unfavorable to the ductility and energy dissipation of the model. This conclusion is in good agreement with experimental results.

In view of the shear wall size and reinforcement involved in this test, it is suggested that the design parameters of EHSW are: the number of prestressed tendons is 3; the pre-tension should be 416.6 kN; the unbonded length of grouted bar should be 150 mm [9].

References

1. Zhuang Z. ABAQUS nonlinear finite element analysis and examples [M]. Beijing: Science Press, 2005.
2. Dassault Systemes. ABAQUS Analysis user's manual version 6.14 [M]. Waltham, MA, 2014.
3. Guo ZH, Shi XD. Principles and analysis of reinforced concrete [M]. Beijing: Tsinghua University Press, 2003.
4. Zhang J, Wang QY, Hu SY, et al. Validation of ABAQUS concrete damaged plasticity model parameters [J]. Building Structure, 2008, 38(8): 127–130.
5. Birtel V, Mark P. Parameterized finite element modeling of RC. ABAQUS Users' Conference [C], 2006.
6. Lu XZ, Ye LP, Miao ZW, et al. Seismic elastoplastic analysis of structures: Principles, models and practices in ABAQUS, MSC.MARC and SAP2000 [M]. Beijing: China Architecture & Building Press, 2009.
7. National standard of the People's Republic of China. Code for design of concrete structures GB 50010–2010 (2015 Edition) [S]. Beijing: China Architecture & Building Press, 2015.
8. Tang L. Research on New hybrid assembly technology for precast concrete shear wall structures [D]. Nanjing: Southeast University, 2016.
9. Zhu ZF, Guo ZX, Tang L. Experimental study and finite element analysis on seismic behavior of new hybrid assembled concrete shear wall [J]. Journal of Civil Engineering, 2018, 51(03): 41–48.

Chapter 5
Design Method of Emulative Hybrid Assembled Precast Concrete Shear Wall

5.1 EHSW Cracking Load Calculation

The test results reveal that the load–displacement curve of the EHSW specimen is linear before concrete cracking, and the specimen can be approximated as being in an elastic working state. At the same time, due to the influence of concrete discontinuity, the initial cracking of EHSW specimens generally occurs at the top of the slurry layer. In this section, the tensile strength contribution of the concrete material to the component's crack resistance is replaced by the bonding force between the grouting material and the concrete interface. This bonding force is generally less than the tensile strength of the concrete material.

The 7.2.3 of the Code for design of concrete structures (GB 50010–2010) [1] provides the calculation method for the anti-cracking bending moment of cast-in-place concrete, which is still used here. Among them, since the cohesive force between the grouting material and the concrete at the joint interface cannot be determined specifically, the calculated value of the concrete tensile strength is still used for calculation.

The theoretical calculation method of horizontal cracking load for EHSW specimens is shown in Eqs. 5.1–5.3.

$$\sigma = \frac{N + P}{A_w} \tag{5.1}$$

$$M_{cr} = \frac{1}{6}bh^2 \cdot (\sigma + \gamma f_t) \tag{5.2}$$

$$F_{cr} = M_{cr}/H \tag{5.3}$$

Z. Zhu et al., *Research on Emulative Hybrid Assembled Precast Concrete Shear Walls*,
https://doi.org/10.1007/978-981-96-9578-2_5

In the formula:

N The vertical force comprises two parts: one is the weight of the model excluding the base, with a theoretical value of 32.9 kN; the other is external axial pressure, which can be taken as 946.3 kN or 473.15 kN depending on the desired axial compression ratio.

P The internal force of prestressed tendons after applying a vertical load is converted based on the measured stress value of prestressed tendons after axial compression, as listed in Table 4.2.

A_w model wall cross-sectional area, calculated 340,000 mm^2,

b model wall section width, take 200 mm,

h model wall section height, take 1700 mm,

$\gamma.$ The plastic influence coefficient of the moment resistance of the section of the member is calculated to be 1.20, in accordance with Clause 7.2.4 of the "Code for design of concrete structures (GB 50010–2010)" [1].

f_t The measured tensile strength of concrete is 2.5 MPa according to .

H The vertical distance from the horizontal load loading point to the horizontal joint section is 3460 mm,

M_{cr} Model wall section cracking moment value.

The comparison between theoretical and measured cracking load values for each EHSW specimen, calculated using Formula 5.1, is presented in Table 5.1.

By comparing the results in Table 5.1, it can be seen that there is a significant difference between the theoretical and measured values, but the theoretical value is less than the measured value, indicating that the formula specified and the interface bonding force based on the tensile strength of concrete still has sufficient safety margin.

The test results and the conclusion of finite element parameter analysis indicate that the pretension and axial compression ratio of prestressed tendons are sensitive parameters that affect the cracking load of EHSW specimens. Based on the theoretical calculation method, specifically Eqs. 5.1–5.3, it is necessary to modify the calculation

Table 5.1 Comparison between theoretical and measured cracking load of EHSW specimen

Test piece	Theoretical value/kN	Measured value/kN	Error (%)[a]	Remark
EHSW1	203	300	− 32	EHSW benchmark model
EHSW2	189	250	− 24	Pretension unchanged, the number of prestressed tendon changes
EHSW3	192	250	− 23	
EHSW4	194	275	− 29	Prestressed tendon number unchanged, prestressed change
EHSW5	187	250	− 25	
EHSW6	201	300	− 33	Variation of unbonded length of grouted anchor bar
EHSW7	204	300	− 32	
EHSW8	165	225	− 27	Axial compression ratio change

Remark [a] Error = (Theoretical value − Measured value)/Measured value × 100%

method of the cracking load of the EHSW specimen to obtain a suitable calculation method that can be essentially consistent with the experimentally measured values. During the correction process, the pre-tension parameters of prestressed tendons are replaced by the comprehensive reinforcement characteristic value ξ_p [2]. The calculation is shown in Eq. 5.4, and the calculation results of the comprehensive reinforcement characteristic value ξ_p of each model are shown in Table 5.2.

$$\xi_p = \frac{\sigma_{pe}A_p + f_yA_s}{f_cbh_p} \tag{5.4}$$

In the formula:

- σ_{pe} Effective prestress of prestressed tendon, according to initial prestress value;
- A_p Prestressed reinforcement area, calculated according to the number of prestressed reinforcement;
- f_y Tensile yield strength of ordinary steel, according to values;
- A_s Tensile ordinary steel bar area, for the shear wall, for the calculation is simple, here approximately take half of the shear wall section reinforcement calculation;
- f_c The compressive strength of concrete, according to axial compressive strength value;
- b Model wall section width, take 200 mm;
- h_p The distance from the action point of the resultant force of the prestressed tendon to the compression edge of the section is 850 mm.

The axial compression ratio of each model is based on the test axial compression ratio, with the specific table being Table 3.1.

Based on the calculation results of the comprehensive reinforcement characteristic value and the axial compression ratio of the model, taking the comprehensive reinforcement characteristic value ξ_p and the axial compression ratio λ as parameters, the theoretical value is fitted to the experimental value using the multiple linear fitting method, and the formula for calculating the cracking load after fitting is given in Eq. 5.5.

$$F'_{cr} = \left(0.078 + 3.512\xi_p + 1.016\lambda\right)F_{cr} \tag{5.5}$$

The cracking load for each model is recalculated using Eq. 5.5, and the results are compared with the test values as shown in Table 5.3. As can be seen from the table, the calculation accuracy is obviously improved after correction, and the error is controlled within − 3% ~ + 5%, which is in good agreement with the measured value.

Table 5.2 The comprehensive reinforcement characteristic value of the EHSW specimen

Test piece	EHSW1	EHSW2	EHSW3	EHSW4	EHSW5	EHSW6	EHSW7	EHSW8
ξ_p	0.33	0.29	0.30	0.30	0.28	0.32	0.33	0.33

Table 5.3 Cracking load correction value of EHSW specimen

Test piece	Corrected value/kN	Measured value/kN	Error (%)[a]
EHSW1	301	300	0.33
EHSW2	252	250	0.80
EHSW3	262	250	4.80
EHSW4	268	275	– 2.55
EHSW5	245	250	– 2.00
EHSW6	293	300	– 2.33
EHSW7	304	300	1.33
EHSW8	225	225	0

Remark [a] Error = (Corrected value – Measured value)/Measured value * 100%

5.2 EHSW Yield Load Calculation

The failure modes of EHSW specimens are all flexural-shear failure, and the root section of the wall is the most unfavorable section, Therefore, the yield load of the member can be determined according to the flexural and shear bearing capacity of the root section of the wall.

5.2.1 EHSW Bending Yield Load

1) Prestressed tendon stress calculation method

To calculate the EHSW bending yield load, the yield strength of ordinary steel bars can be measured. However, the actual stress of unbonded prestressed tendons at this stage needs to be determined by calculation.

The calculation of ultimate stress or stress increment for unbonded prestressed tendons is currently limited to beam and plate members, with a variety of calculation methods having been developed. These methods can be divided, based on their calculation principles, into bond reduction factor method, regression formula based on section reinforcement index, and calculation method based on deformation or equivalent plastic hinge length. Representative calculation methods include the ACI 318 specification recommended formula [3], Du Gongchen and Tao Xuekang formula [4], and the "Technical specification for concrete structures prestressed unbonded tendons (JGJ 92–2016)" [5] recommended formula.

(1) ACI 318 Recommended Formula

The calculation method for ultimate stress in unbonded prestressed tendons f_{ps} is specified in Article 18.7.2 of ACI 318-11. The calculation formula is shown in Formula 5.6, which converts the original formula into units. For the case of this specimen, the span-depth ratio is not greater than 35.

$$f_{ps} = f_{se} + 70\,\text{MPa} + \frac{f_c'}{100\rho_p} \tag{5.6}$$

In the formula:

f_{se} Effective prestress of prestressed tendon;
f_c' Concrete compressive strength;
ρ_p Prestressed reinforcement ratio, calculated by A_p/bd_p,
A_p is the prestressed reinforcement area,
b is the beam section compression flange width,
d_p is the prestressed reinforcement center of gravity to the section compression edge distance.

(2) Du Gongchen and Tao Xuekang Formula

Du Gongchen and Tao Xuekang introduced the comprehensive reinforcement index ξ_p, and established a linear relationship between ξ_p and the ultimate stress of unbonded prestressed tendons based on experimental data, as shown in Eq. 5.7, where the calculation of ξ_p is the same as Eq. 5.4.

$$f_{ps} = f_{pe} + \left(786 - 1920\xi_p\right)\text{MPa} \tag{5.7}$$

(3) Recommended formula for *Technical specification for concrete structures prestressed with unbonded tendons JGJ 92 - 2016*

In *Article 5.1.11 of JGJ 92-2016*, the calculation method of stress design value (ultimate stress value) f_{ps} of unbonded prestressed tendons in flexural members using unbonded prestressed tendons is given, as shown in Eq. 5.8.While considering the comprehensive reinforcement index (the characteristic value of comprehensive reinforcement), the influence of height-span ratio $\frac{h}{l_0}$ of members is introduced.

$$f_{ps} = f_{pe} + \left(240 - 335\xi_p\right)\left(0.45 + 5.5\frac{h}{l_0}\right) \tag{5.8}$$

Due to the absence of a calculation method for the ultimate stress of unbonded prestressed tendons in shear wall members, this book selects the aforementioned three methods for trial calculation, and compares the calculated values with experimental values to identify the most accurate method as the ultimate stress calculation method for EHSW unbonded prestressed tendons. In the calculation process, according to Table 4.2, the initial prestress value is taken from Table 3.2, the axial compressive strength value is taken from Table 5.2, and the value of interest $\frac{h}{l_0}$ is 1700/3460 = 0.49.T The results calculated by each method are presented in Tables 5.4, 5.5 and 5.6.

The results in Tables 5.4, 5.5 and 5.6 show that the JGJ 92 method overestimates the ultimate stress of unbonded prestressed tendons, while the ACI 318 formula generally underestimates it. Therefore, in this book, the ultimate stress of unbonded

Table 5.4 EHSW calculation of ultimate stress of unbonded prestressed tendons (ACI 318 method)

Test piece	Measured value/kN	Theoretical value/kN	Error (%)[a]
EHSW1	1158.9	1022.2	– 11.80
EHSW2	1082.1	924.4	– 14.57
EHSW3	1416.1	1480.0	4.51
EHSW4	948.2	818.7	– 13.66
EHSW5	808.9	672.2	– 16.90
EHSW6	1127.6	969.5	– 14.02
EHSW7	1177.6	1035.6	– 12.06
EHSW8	1167.4	1027.1	– 12.02

Remark [a] Error = (Theoretical value − Measured value)/Measured value × 100%

Table 5.5 Calculation of ultimate stress of unbonded prestressed tendons of EHSW specimen (Du and Tao Method)

Test piece	Measured value/kN	Theoretical value/kN	Error (%)[a]
EHSW1	1158.9	1031.4	– 11.00
EHSW2	1082.1	992.7	– 8.26
EHSW3	1416.1	1483.2	4.74
EHSW4	948.2	882.2	– 6.96
EHSW5	808.9	774.8	– 4.22
EHSW6	1127.6	992.8	– 11.95
EHSW7	1177.6	1041.2	– 11.58
EHSW8	1167.4	1035.0	– 11.34

Remark [a] Error = (Theoretical value − Measured value)/Measured value × 100%

Table 5.6 Calculation of ultimate stress of unbonded prestressed tendons of EHSW specimen (JGJ 92 Method)

Test piece	Measured value/kN	Theoretical value/kN	Error (%)[a]
EHSW1	1158.9	1286.7	11.03
EHSW2	1082.1	1210.5	11.87
EHSW3	1416.1	1708.7	20.66
EHSW4	948.2	1113.0	17.38
EHSW5	808.9	988.0	22.14
EHSW6	1127.6	1241.7	10.12
EHSW7	1177.6	1298.1	10.23
EHSW8	1167.4	1290.9	10.58

Remark [a] Error = (Theoretical value − Measured value)/Measured value × 100%

prestressed tendons is preliminarily calculated by the Du and Tao methods, and the theoretical calculation results are directly used to calculate flexural bearing capacity at this stage.

2) Calculation of bending capacity at yield stage

The EHSW specimen is subjected to both vertical constant and horizontal monotonic loads. The calculation method for eccentric compression members can be utilized to determine the flexural capacity of the normal section. Based on the failure mode of each test specimen, it can be preliminarily judged that the calculation mode of each specimen can be attributed to the large eccentric compression member. Therefore, it is preliminarily assumed that the normal section flexural bearing capacity of the large eccentric compression member is calculated, and the relative compression zone height is reviewed in the calculation process. If not satisfied, the relevant calculation formula for the small eccentric compression member is recalculated. Following the calculation method for large eccentric compression members and considering the contribution of unbonded prestressed tendons to bearing capacity, the basic calculation process is presented in Eq. 5.9 through Eq. 5.12.

$$N = \alpha_1 f_c bx + A'_s f_y - A_s f_y - (h_{w0} - 1.5x)\frac{A_{sw}}{h_{w0}} f_{yw} - A_p f_{ps} \tag{5.9}$$

$$N\left(e_0 - \frac{h_w}{2} + \frac{x}{2}\right) = A_s f_y\left(h_{w0} - \frac{x}{2}\right) + A'_s f_y\left(\frac{x}{2} - a\right) + A_p f_{ps}\left(h_p - \frac{x}{2}\right) + (h_{w0} - 1.5x)\frac{A_{sw} f_{yw}}{h_{w0}}\left(\frac{h_{w0}}{2} + \frac{x}{4}\right) \tag{5.10}$$

$$h_{w0} = h_w - a_s \tag{5.11}$$

$$M_y = Ne_0 \tag{5.12}$$

In the formula:

N The vertical load of the model is the same as that of Formula 5.1, and 979.2 kN and 506.05 kN are taken respectively according to different axial compression ratios.

α_1 The ratio of the rectangular stress diagram stress to the axial compressive strength of concrete in the compression zone is related to the concrete strength grade. For EHSW specimen, the concrete strength grade is C35, not more than C50, and the value is 1.0.

f_c The axial compressive strength of concrete is 23.7 MPa according to .

A'_s The vertical reinforcement area of the edge member of the compression side of the wall limb is 1608 mm^2 for the EHSW specimen;

A_s The vertical reinforcement area of the edge member of the tension side of the wall limb is 1608 mm^2 for the EHSW specimen;

f_y The yield strength of the tensile steel bar of the edge member is measured here, and the value is 438 MPa according to .
h_w Wall section height, 1700 mm;
a_s The distance from the center of gravity of the vertical steel bar to the edge of the tensile side of the member is 200 mm;
h_{w0} The effective height of the wall section is calculated to be 1500 mm;
A_{sw} The total area of the vertical distribution of steel bars in the middle of the wall limb is 628 mm^2 for the EHSW specimen.
f_{yw} The yield strength of the steel bar is distributed vertically, and the measured yield strength of the steel bar is taken here, and the value is 430 MPa according to .
A_p Unbonded prestressed tendon area, depending on the number, can be taken as 560 mm^2, 420 mm^2, 280 mm^2;
f_{ps} Unbonded prestressed tendon limit stress value, according to Table 5.5 take the corresponding theoretical value;
h_p the distance from the center of gravity of unbonded prestressed tendon to the edge of compressed concrete is 850 mm;
e_0 Equivalent eccentricity of vertical load;
M_y The flexural capacity of the normal section corresponding to N.

In the case of symmetrical reinforcement, due to $A'_s = A_s$, Formula 5.9 can be simplified and deduced, the calculation formula of relative compression zone height ξ is obtained, see Formula 5.13.

$$\xi = \frac{x}{h_{wo}} = \frac{N + A_{sw}f_{yw} + A_p f_{ps}}{\alpha_1 f_c b h_{w0} + 1.5 A_{sw} f_{yw}} \tag{5.13}$$

Expand Eq. 5.10 and substitute it into Eq. 5.12, the M_y calculation formula can be obtained, see Eq. 5.14.

$$M_y = N\left(\frac{h_w}{2} - \frac{x}{2}\right) + A_s f_y (h_{w0} - a) + A_p f_{ps}\left(h_p - \frac{x}{2}\right) + (h_{w0} - 1.5x)\frac{A_{sw} f_{yw}}{h_{w0}}\left(\frac{h_{w0}}{2} + \frac{x}{4}\right) \tag{5.14}$$

According to the calculated M_y, the corresponding horizontal load F_y at the top of the wall limb can be obtained, see Formula 5.15.

$$F_y = M_y / H_w \tag{5.15}$$

In the formula

H_w The height of the horizontal load action point to the root section of the wall limb is 3460 mm.

Table 5.7 Theoretical value of bending yield load of EHSW specimen

Test piece	ξ	x/mm	M_y/kN m	F_y/kN
EHSW1	0.243	364.63	2099	606.8
EHSW2	0.222	332.57	2020	583.9
EHSW3	0.221	332.24	2019	583.7
EHSW4	0.232	347.96	2059	595.1
EHSW5	0.224	335.95	2029	586.4
EHSW6	0.240	360.32	2089	603.8
EHSW7	0.244	365.73	2102	607.5
EHSW8	0.180	270.59	1853	535.4

The bending yield load of each EHSW specimen is calculated according to Eqs. 5.9–5.15, and the results are shown in Table 5.7.

5.2.2 *EHSW Shear Yield Load*

For EHSW specimens, overall slip along the joint may still occur, generally described by the friction shear mechanism. The corresponding theoretical calculation formula is 5.16, safely ignoring the beneficial effect of stress increment of unbonded prestressed tendons.

$$V_y = 0.6 f_y A_s + 0.8\,\mathrm{N} \tag{5.16}$$

In the formula:

V_y Shear yield load under overall slippage of horizontal abutments;
f_y Through the horizontal seam ordinary steel yield strength, according to measured yield strength;
A_s Ordinary steel area through horizontal seam.

The shear yield load of each EHSW specimen is calculated according to Eq. 5.16. The results are shown in Table 5.8.

Table 5.8 Theoretical value of shear yield load of EHSW specimen

Test piece	EHSW1	EHSW2	EHSW3	EHSW4	EHSW5	EHSW6	EHSW7	EHSW8
V_y/kN	1790.5	1790.5	1790.5	1790.5	1790.5	1790.5	1790.5	1412.0

5.2.3 EHSW Yield Load Determination

Comparing the theoretical values of horizontal load in Tables 5.7 and 5.8, it can be found that the horizontal load calculated by shear yield is much higher than the horizontal load determined by bending yield, indicating that the EHSW specimen is controlled by bending yield, which is also consistent with the experimental phenomenon. Therefore, the yield load of EHSW should be determined based on the bending yield state, specifically Table 5.7, and the calculation results should be compared with the measured values, as shown in Table 5.9. Among them, the measured value of the test in the table is different from the yield load determined by the yield of the first tensile steel bar during the test, but the horizontal load corresponding to the yield displacement determined by Fig. 3.27 is taken and the positive direction is taken.

As can be seen from Table 5.9, the theoretical value is generally lower than the measured value, indicating that the yield load of EHSW calculated by the existing theoretical method can yield a more conservative result. The analysis indicates that the difference between the theoretical and measured values is primarily due to the discreteness of the actual material performance, and the approximation of the stress increment calculation method for unbonded prestressed tendons based on beam and slab flexural members applied to shear wall members; there must be many factors such as errors in the test loading process.

The test results and the conclusion of finite element parameter analysis indicate that the area of prestressed reinforcement, the pretension of prestressed reinforcement, the unbonded length of slurry anchor reinforcement, and the axial compression ratio are all sensitive parameters affecting the yield load of EHSW specimens, and are positively correlated with each parameter. Based on the theoretical calculation method, i.e., Eq. 5.7, it is necessary to modify the calculation method of the cracking load of the EHSW specimen to obtain a suitable calculation method that can be basically consistent with the experimental measured values. In the correction process,

Table 5.9 Comparison between theoretical and measured yield load of EHSW specimen

Test piece	Theoretical value/kN	Measured value/kN	Error (%)[a]	Remark
EHSW1	606.8	673.3	– 9.88	EHSW benchmark model
EHSW2	583.9	681.3	– 14.30	Pretension unchanged, the number of prestressed tendon changes
EHSW3	583.7	646.3	– 9.69	
EHSW4	595.1	681.3	– 12.65	Prestressed tendon number unchanged, prestressed change
EHSW5	586.4	689.3	– 14.93	
EHSW6	603.8	648.2	– 6.85	Variation of unbonded length of grouted anchor bar
EHSW7	607.5	685.3	– 11.35	
EHSW8	535.4	584.9	– 8.46	Axial compression ratio change

Remark [a] Error = (Theoretical value – Measured value)/Measured value × 100%

the area parameter of prestressed reinforcement is expressed by the reinforcement ratio of prestressed reinforcement ρ_p (see Eq. 5.6), the pre-tension parameter of prestressed reinforcement is expressed by the characteristic value of comprehensive reinforcement ξ_p (see Eq. 5.4), and the unbonded length parameter of grouted anchor steel bar is expressed by the ratio of the unbonded length of grouted anchor steel bar to the nominal diameter of steel bar n. The above-mentioned three types of parameters for EHSW specimens are presented in Table 5.10.

The multiple linear fitting method is utilized to fit the theoretical value to the experimental data, and the yield load calculation formula after fitting is 5.17.

$$F_y' = \left(1.034 + 2.118\rho_p + 0.001\xi_p + 0.001n + 0.693\lambda\right)F_y \tag{5.17}$$

The yield load of each model is recalculated according to Eq. 5.17, and the comparison with the measured values is shown in Table 5.11. As can be seen from the table, calculation accuracy is obviously improved after correction, and the error is controlled within − 4% to + 6%, in good agreement with the measured value.

Table 5.10 EHSW specimen yield load related parameters

Test piece	ρ_p	ξ_p	n	λ
EHSW1	0.00329	0.33	12.5	0.11
EHSW2	0.00247	0.29	12.5	0.11
EHSW3	0.00165	0.30	12.5	0.11
EHSW4	0.00329	0.30	12.5	0.11
EHSW5	0.00329	0.28	12.5	0.11
EHSW6	0.00329	0.32	15.625	0.11
EHSW7	0.00329	0.33	9.375	0.11
EHSW8	0.00329	0.33	12.5	0.055

Table 5.11 The yield load correction value of EHSW specimen

Test piece	Corrected value/kN	Measured value/kN	Error (%)[a]
EHSW1	685.7	673.3	1.84
EHSW2	658.8	681.3	− 3.30
EHSW3	657.5	646.3	1.73
EHSW4	672.5	681.3	− 1.29
EHSW5	662.6	689.3	− 3.87
EHSW6	684.2	648.2	5.55
EHSW7	684.6	685.3	− 0.10
EHSW8	584.6	584.9	− 0.05

Remark [a] Error = (Corrected value − Measured value)/Measured value × 100%

5.3 EHSW Limit Load Calculation

The ultimate load calculation method and process is the same as that for yield load in Sect. 5.2, with the consideration of steel bar material strain strengthening. Specifically, the ultimate strength of steel bars should be used instead of yield strength. The specific calculation process is omitted here; only key calculation results are given (see Tables 5.12, 5.13 and 5.14).

As can be seen from Tables 5.12 and 5.13, the calculation of the ultimate load of EHSW is still dominated by bending control, which is also consistent with the experimental phenomenon.

As shown in Table 5.14, similar to the calculation of yield load, the theoretical calculation value of ultimate load for the EHSW specimen is lower than the experimental measured value, indicating that the calculation of ultimate load for EHSW using the existing theoretical method can result in more conservative estimates. It should be noted that some EHSW specimens did not exhibit a significant decrease in bearing capacity at the end of the test, resulting in the ultimate load appearing at the end of the test. For establishing the subsequent restoring force model, considering that the bearing capacity of most models is reduced to a certain extent, the value of the ultimate load here is taken from the measured value of the positive horizontal

Table 5.12 The theoretical value of bending ultimate load of EHSW specimen

Test piece	ξ	x/mm	M_u/kN.m	F_u/kN
EHSW1	0.250	374.59	2512	726.1
EHSW2	0.229	343.03	2435	703.9
EHSW3	0.228	342.70	2435	703.6
EHSW4	0.239	358.17	2473	714.7
EHSW5	0.231	346.36	2444	706.3
EHSW6	0.247	370.34	2502	723.2
EHSW7	0.250	375.67	2515	726.8
EHSW8	0.188	282.02	2272	656.6

Table 5.13 Theoretical value of shear limit load of EHSW specimen

Test piece	f_u/E_s	f_{ps}/MPa	V_u/kN
EHSW1	0.0031	1484.8	2189.6
EHSW2	0.0031	1363	2189.6
EHSW3	0.0031	1870.6	2189.6
EHSW4	0.0031	1281.3	2189.6
EHSW5	0.0031	1134.8	2189.6
EHSW6	0.0031	1432.1	2189.6
EHSW7	0.0031	1498.2	2189.6
EHSW8	0.0031	1489.7	1811.1

Table 5.14 Comparison between theoretical value and measured value of ultimate load of EHSW specimen

Test piece	Theoretical value/kN	Measured value/kN	Error (%)[a]	Remark
EHSW1	726.1	805	– 9.80	EHSW benchmark model
EHSW2	703.9	817	– 13.84	Pretension unchanged, the number of prestressed tendon changes
EHSW3	703.6	793	– 11.27	
EHSW4	714.7	836	– 14.51	Prestressed tendon number unchanged, prestressed change
EHSW5	706.3	784	– 9.91	
EHSW6	723.2	793	– 8.80	Variation of unbonded length of grouted anchor bar
EHSW7	726.8	864	– 15.88	
EHSW8	656.6	759	– 13.49	Axial compression ratio change

Remark [a] Error = (Theoretical value – Measured value)/Measured value × 100%

load at the previous stage of failure of each EHSW specimen, which is different from the peak load term given in Table 3.13.

Similarly, to obtain a method for calculating the ultimate load of EHSW specimens that is consistent with measured values, it is necessary to modify existing theoretical formulas. The test results and finite element parameter analysis indicate that the parameters most influencing the ultimate load of EHSW specimens are the same as those affecting yield load. These include the area of prestressed tendons, the pretension of prestressed tendons, the unbonded length of grouted anchor bars, and the axial compression ratio. In addition to the negative correlation with the unbonded length of grouted anchor bars, it is positively correlated with other parameters. Therefore, the impact of the above three types of parameters is introduced in the correction process, and the parameters are assigned according to Table 5.10.

The theoretical value is fitted to the experimental value using a multiple linear fitting method, and the ultimate load calculation formula is given in Eq. 5.18.

$$F'_u = \left(1.298 + 0.432\rho_p + 0.067\xi_p - 0.014n + 0.001\lambda\right)F_u \tag{5.18}$$

The ultimate load of each model is recalculated according to Formula 5.18, and the comparison with the measured values is shown in Table 5.15. As can be seen from the table, the calculation accuracy is obviously improved after correction, and the error is controlled within – 2% ~ + 6%, which is in good agreement with the measured value.

Table 5.15 Ultimate load correction value of EHSW specimen

Test piece	Corrected value/kN	Measured value/kN	Error (%)[a]
EHSW1	832.6	839	0.77
EHSW2	804.9	816	1.38
EHSW3	804.9	792	– 1.60
EHSW4	818.1	836	2.19
EHSW5	807.5	851	5.39
EHSW6	797.1	796	– 0.14
EHSW7	865.2	864	– 0.14
EHSW8	752.8	759	0.82

Remark [a] Error = (Corrected value − Measured value)/Measured value × 100%

5.4 EHSW Stiffness Calculation

To gain a deeper understanding of the deformation performance of EHSW specimens, a discussion on the stiffness calculation method is presented. With the increase of horizontal load, due to the influence of concrete cracking, steel bar yielding, and other factors, the specimen transitions from elasticity to elastoplastic working state, and its lateral stiffness gradually degrades. To understand the entire process of stiffness and deformation development, the stiffness of the EHSW specimen in the cracking stage, yield stage, and ultimate stage is calculated theoretically, aiming to derive a suitable stiffness calculation method that accurately reflects the experimental conditions.

5.4.1 EHSW Cracking Stage Stiffness Calculation

The stiffness calculation for the cracking stage of the specimen can be assumed to occur when the specimen is still in the elastic stage. Based on the analysis of horizontal concentrated load acting on the vertex of the wall limb, and considering both the bending and shear deformations of the wall limb, the calculation method for lateral stiffness K_{cr} in the cracking stage is 5.19.

$$K_{cr} = \frac{1}{\frac{H_w^3}{3EI} + \frac{\mu H_w}{GA}} \tag{5.19}$$

In the formula:

H_w The height of the model is calculated, where the height of the horizontal load point to the horizontal joint section is 3460 mm;

E Elastic modulus of concrete material, according to take 31444 N/mm^2;

I Wall limb section moment of inertia, $I = bh_w^3/12$, according to the wall limb section size, calculated 81,883,333,333 mm^4;

μ Shear stress non-uniformity coefficient, for rectangular section, take 1.2;
G The shear deformation modulus of the concrete material is 40% of the elastic modulus, that is, 12,577.6 N/mm^2;
A Wall section area, $A = bh_w$, according to the wall section size, calculated 340,000 mm^2.

According to the lateral stiffness K_{cr} of the EHSW specimen in the cracking stage calculated by Eq. 5.19, and the correction value of the cracking load calculated by Eq. 5.1 (see Eqs. 5.1–5.5), the theoretical solution of the cracking displacement is calculated. The relevant calculation results are shown in Table 5.16.

Table 5.16 shows that the theoretical value of cracking displacement differs significantly from the measured value, and is much less than the measured cracking displacement. The analysis shows that the main reason is that Formula 5.19 does not consider the material micro-damage that must exist in the cracking process of the model, but takes the theoretical elastic value of the material performance parameters. This overestimates the stiffness of the EHSW specimen at this stage. Hence, it is necessary to introduce a proportional coefficient α based on Formula 5.19 to properly consider the reduction effect of material micro-damage on the overall stiffness of the model.

Furthermore, based on the test results and finite element parameter analysis, the parameters with significant influence on the cracking stiffness of EHSW specimens include the pre-tension of prestressed tendons and the axial compression ratio, which are positively correlated. Similar to the above, Formula 5.19 is improved by introducing scale factor α, comprehensive reinforcement characteristic value ξ_p and axial compression ratio λ, and Formula 5.20 is constructed. Multiple linear regressions are performed, and the theoretical displacement in the cracking stage is approximated to the measured value. Finally, the applicable stiffness correction formula in

Table 5.16 Comparison between theoretical and measured cracking displacement of EHSW specimen

Test piece	K_{cr}/(kN/mm)	Theoretical value of cracking displacement/mm	Cracking displacement measured value/mm	Remark
EHSW1	157.89	1.90	6.80	EHSW benchmark model
EHSW2	157.89	1.61	5.47	Pretension unchanged, the number of prestressed tendon changes
EHSW3	157.89	1.67	5.31	
EHSW4	157.89	1.71	5.75	Prestressed tendon number unchanged, prestressed change
EHSW5	157.89	1.57	5.82	
EHSW6	157.89	1.86	7.37	Variation of unbonded length of grouted anchor bar
EHSW7	157.89	1.92	6.86	
EHSW8	157.89	1.42	6.09	Axial compression ratio change

Table 5.17 Comparison of modified and measured cracking displacement of EHSW specimen

Test piece	K'_{cr}/(kN/mm)	Cracking displacement correction value/mm	Cracking displacement measured value/mm	Error (%)[a]
EHSW1	44.87	6.75	6.80	– 0.74
EHSW2	44.86	5.65	5.47	3.29
EHSW3	44.87	5.88	5.31	10.73
EHSW4	44.87	6.01	5.75	4.52
EHSW5	44.86	5.50	5.82	– 5.50
EHSW6	44.87	6.57	7.37	– 10.85
EHSW7	44.87	6.82	6.86	– 0.58
EHSW8	36.85	6.10	6.09	0.16

Remark [a] Error = (Cracking displacement correction value − Cracking displacement measured value)/Cracking displacement measured value × 100%

the cracking stage is obtained, which is consistent with the test data, as shown in Eq. 5.21.

$$K'_{cr} = \alpha f\left(\xi_p, \lambda\right) K_{cr} \tag{5.20}$$

In the formula:

$f\left(\xi_p, \lambda\right)$ A multivariate linear polynomial containing ξ_p, λ.

$$K'_{cr} = 0.542\left(0.355 + 0.001\xi_p + 1.381\lambda\right) K_{cr} \tag{5.21}$$

$$\Delta'_{cr} = F'_{cr}/K'_{cr} \tag{5.22}$$

The lateral stiffness and corresponding horizontal displacement of each EHSW specimen at the cracking stage were recalculated according to Eqs. 5.21–5.22, and compared with the measured values. The calculation results are presented in Table 5.17. As can be seen from the table, the accuracy of the predicted displacement measured values has significantly improved. Overall, Eq. 5.21 provides a reasonable prediction of the stiffness of the EHSW specimen at the cracking stage.

5.4.2 *EHSW Yield Stage Stiffness Calculation*

In the yield stage of EHSW specimen, the concrete has been damaged to a certain extent and the vertical steel bar has yielded, and the model has entered the elastoplastic working stage. Generally, the calculation is simplified, and the stiffness of this stage is generally taken as 85% of the stiffness of the cracking stage, that is, the elastic stage.

Furthermore, based on the test results and the conclusion of finite element parameter analysis, the parameters that exhibit a significant influence on the yield stiffness of EHSW specimens and adhere to clear rules are the pre-tension of prestressed tendons, the unbonded length of grouted anchor bars, and the axial compression ratio. These parameters are positively correlated with the pre-tension of prestressed tendons and the axial compression ratio, while being negatively correlated with the unbonded length of grouted anchor bars. As in the previous article, the comprehensive reinforcement characteristic value ξ_p, the ratio of the unbonded length of the slurry anchor steel bar to the diameter of the steel bar n, and the axial compression ratio λ are introduced, and the structural Formula 5.23 is constructed, and multiple linear fitting is performed. The theoretical value of displacement in the yield stage is approximated to the measured value, and a corresponding yield stage stiffness correction formula is derived, see Formula 5.24.

$$K_y' = 0.85f\left(\xi_p, n, \lambda\right)K_{cr}' \tag{5.23}$$

In the formula:

$f\left(\xi_p, n, \lambda\right)$ Multivariate linear polynomials containing ξ_p, n, λ.

$$K_y' = 0.85\left(0.587 + 0.001\xi_p - 0.011n + 0.001\lambda\right)K_{cr}' \tag{5.24}$$

$$\Delta_y' = \Delta_{cr}' + \left(F_y' - F_{cr}'\right)/K_y' \tag{5.25}$$

By Eqs. 5.24 and 5.25, the lateral stiffness and corresponding horizontal displacement of each EHSW specimen at yield are calculated and compared to measured values. The results are shown in Table 5.18. Among them, the yield displacement measured value according to Table 3.15 is a positive yield displacement value. The analysis shows that the linear correlation between the change of the axial compression ratio parameter and the change of the yield displacement measured value, as inferred from the measured data of yield displacement, cannot be fully applied. The analysis indicates that the linear correlation between the change of the axial compression ratio parameter and the change of the yield displacement measured value, as inferred from the measured data of yield displacement, cannot be fully applied. Therefore, there must be some errors in the linear fitting of this book.

Table 5.18 Comparison between corrected and measured yield displacement of EHSW specimen

Model number	K_y'/(kN/mm)	Yield displacement correction value/mm	Measured yield displacement/mm	Error (%)[a]
EHSW1	17.16	29.2	31.2	– 6.41
EHSW2	17.16	29.4	28.5	3.16
EHSW3	17.16	28.9	27.5	5.09
EHSW4	17.16	29.6	30.0	– 1.33
EHSW5	17.15	29.8	32.4	– 8.02
EHSW6	15.85	31.3	31.6	– 0.95
EHSW7	18.47	27.4	29.1	– 5.84
EHSW8	14.09	31.6	28.4	11.27

Remark [a] Error = (v-Measured yield displacement)/Measured yield displacement × 100%

5.4.3 EHSW Limit Stage Stiffness Calculation

The stiffness calculation of EHSW specimens in the limit stage is essentially the same as that in the yield stage. For simplified calculation of the stiffness of EHSW specimens in the limit stage, Formula 5.26 is constructed; however, the reduction coefficient of stiffness in this stage compared to the cracking stage, i.e., the elastic stage β, is determined based on statistical analysis of test data. At the same time, considering that the elastic–plastic development of the limit stage model is more sufficient, the relevant design parameters have a certain influence on its stiffness. Therefore, this study considers the influence of the prestressed reinforcement ratio ρ_p, the comprehensive reinforcement characteristic value ξ_p, the ratio of the unbonded length of the slurry anchor steel bar to the diameter of the steel bar n, and the axial compression ratio λ. Multivariate linear fitting is performed. The theoretical value of the limit stage displacement is approximated to the measured value, and finally an applicable and consistent result with the test data is obtained. The limit stage stiffness correction formula, see Formula 5.27.

$$K_u' = \beta f\left(\rho_p, \xi_p, n, \lambda\right)K_{cr}' \tag{5.26}$$

In the formula

$f\left(\rho_p, \xi_p, n, \lambda\right)$ Multivariate linear polynomials containing ρ_p, ξ_p, n, λ.

$$K_u' = 0.384\left(0.205 + 1.595\rho_p + 0.482\xi_p - 0.012n + 0.001\lambda\right)K_{cr}' \tag{5.27}$$

$$\Delta_u' = \Delta_y' + \left(F_u' - F_y'\right)/K_u' \tag{5.28}$$

Table 5.19 Comparison between corrected and measured ultimate displacement of EHSW specimen

Model number	K'_u/(kN/mm)	The limit displacement correction value/mm	The limit displacement measured value/mm	Error (%)[a]
EHSW1	3.78	68	70.5	– 3.55
EHSW2	3.43	72	70.2	2.56
EHSW3	3.49	71.2	70.3	1.28
EHSW4	3.53	70.8	70.6	0.28
EHSW5	3.36	72.9	70.6	3.26
EHSW6	3.05	68.3	70.7	– 3.39
EHSW7	4.43	68.2	70.6	– 3.40
EHSW8	3.10	85.8	70.5	21.70

Remark [a] Error = (The limit displacement correction value − The limit displacement measured value)/The limit displacement measured value × 100%

According to Formulas 5.27 and 5.28, the lateral stiffness and corresponding horizontal displacement of each EHSW specimen in the ultimate stage are calculated and compared with the measured values. The calculation results are shown in Table 5.19. Among them, for the value of the measured displacement in the limit stage, the measured positive horizontal displacement corresponding to the 5.3 ultimate load is taken. Table 5.19 shows that Eq. 5.27 provides a good prediction of EHSW stiffness at this stage, but like the yield stage, the deviation of the EHSW8 model with small axial compression ratio is still large. It is also believed that from the measured data of ultimate displacement, the linear correlation between the change of axial compression ratio parameters and the measured value of ultimate displacement is still not fully applicable. Therefore, there must be some errors in the linear fitting of this paper.

References

1. National standard of the People's Republic of China. Code for design of concrete structures GB 50010–2010 (2015 Edition). Beijing: China Building Industry Press; 2015.
2. Tang L. Research on new hybrid assembly technology of precast concrete shear wall structure. Nanjing: Southeast University; 2016.
3. ACI Committee 318. ACI 318–11 Building Code Requirements for Structural Concrete and Commentary. American Concrete Institute, U.S.A.; 2011.
4. Du GC, Tao XK. Ultimate stress of unbonded tendons in partially prestressed concrete beams. PCI J. 1985;30(6):72–91.
5. Industry standard of the People's Republic of China. Technical specification for concrete structures prestressed with unbonded tendons JGJ 92–2016 [S]. Beijing: China Architecture & Building Press, 2016.

Chapter 6
Restoring Force Model of Emulative Hybrid Assembled Precast Concrete Shear Wall

When a structure or component undergoes deformation due to external forces, it attempts to revert to its original resistance state, demonstrating its restoring force. The relationship between this force and deformation is depicted by the restoring force characteristic curve, commonly referred to as the hysteresis curve. This curve encapsulates essential mechanical properties such as strength, stiffness, and ductility of the component. The area enclosed by the hysteresis loop indicates the component's energy absorption capacity, which serves as a vital metric in seismic performance analysis.

In the seismic response analysis of structures or components, the actual restoring force characteristic curve is typically represented by a restoring force model. This model is a practical mathematical formulation derived through appropriate abstraction and simplification from numerous force–deformation curves obtained through experimentation. Establishing an effective restoring force model for emulative hybrid assembled precast concrete shear walls (EHSW) forms the foundation for theoretical seismic performance analysis and seismic design, holding considerable theoretical significance.

6.1 Overview of Restoring Force Model

6.1.1 Classification of Restoring Force Models

Research on restoring force models has been conducted in the field of earthquake engineering both domestically and internationally since the 1960s, leading to the proposal of various restoring force models for theoretical analysis and engineering design reference. From the perspective of the analysis object, restoring force models can be categorized into three levels: material restoring force models, component restoring force models, and structural restoring force models. The material restoring

Z. Zhu et al., *Research on Emulative Hybrid Assembled Precast Concrete Shear Walls*,
https://doi.org/10.1007/978-981-96-9578-2_6

force model is grounded in the properties of materials such as steel and concrete, specifically focusing on the stress–strain relationships of these materials. Notably, the steel restoring force model has been primarily developed by researchers including Kato, Zhu Bolong, and Sozen, while the concrete restoring force model encompasses the Darwin and Peeknold models, as well as the models proposed by Lv et al. [1]. The component restoring force model is predicated on individual components, such as reinforced concrete beams, columns, shear walls, and beam-column joints, and is characterized by the bending moment–curvature relationship and the force–displacement relationship. The restoring force models for beams and columns are essential when conducting elastic–plastic seismic response analysis of structures using a bar model. Conversely, when employing a layer model for elastic–plastic seismic response analysis, the equivalent shear stiffness and shear bearing capacity of each layer are determined based on the bending, axial, and shear stiffness and bearing capacity of each component, thereby facilitating the establishment of a layer restoring force model through the static elastic–plastic method.

The determination method of the restoring force model is divided into three categories: experimental fitting method, system identification method, and theoretical calculation method. A certain mathematical model is used to quantify the skeleton curve and hysteresis loops under different control standards from the scatter diagram of test results. Then, the skeleton curve and standard hysteresis loops are combined to form the restoring force curve, from which the stiffness degradation law under repeated loading is determined by comparing the standard hysteresis loops under different control deformation. In the system identification method, three parameters of the restoring force model are summarized according to an existing model: those controlling stiffness degradation, those controlling pinching effect, and those controlling strength degradation. On this basis, dynamic parameters are identified through shaking table tests or calculations. Theoretical calculation method involves calculating and simplifying the restoring force model at the material level to obtain a component level model.

6.1.2 Content to Be Determined for Restoring Force Model

To determine the restoring force model, one must first identify the skeleton curve and hysteresis rules. The skeleton curve helps to pinpoint key points, which in turn reflect major features such as cracking, yielding, and failure; similarly, the hysteresis rule indicates the high nonlinearity of the component. Generally, key points must be identified during both positive and negative loading and unloading processes, denoting strength degradation and stiffness degradation respectively. Once these points are established, a restoring force model can be built. A commonly employed method involves using a more reliable theoretical formula to pinpoint key points on the skeleton curve and conducting a low cyclic loading test to determine the hysteresis law.

6.1.3 Restoring Force Model of Shear Wall Members

Because of the more complex section form, reinforcement structure, and stress mechanism compared to linear members like beams and columns, shear wall members have numerous influencing factors and an unintuitive law. Presently, there are few mature restoring force models for shear wall members; instead, some approximate macro-analysis models are predominantly used to simulate shear wall member restoring forces. These macro-models generally assemble beam, bar, and support components with relatively mature restoring force models to approximately capture hysteretic characteristics, including the equivalent beam model, equivalent support model, three vertical bar element model, and multi-vertical bar element model [2–6]. These macroscopic models, based on the plane section assumption, are not suitable for shear wall members with shear deformations often accounting for a considerable proportion of total deformations.

Due to the large number of applications of shear wall structures in China, some domestic scholars have conducted research on the restoring force model of shear wall members. Li [7] established the trilinear skeleton curve model and restoring force model of high strength concrete-steel composite shear wall and high strength concrete composite shear wall with polystyrene board based on the results of low cyclic loading test. Liu [8] established a simplified three-line skeleton curve and restoring force model of double-layer steel plate infilled concrete composite shear wall with binding bars by experimental fitting method. Li [9] systematically studied T-shaped, one-shaped and L-shaped short-leg shear walls. Based on the experimental data, the four-line and three-line skeleton curve models were established by least square method, and the corresponding restoring force model was established based on this. Bai [10] carried out experimental research on steel reinforced high performance concrete, and established four-fold line and three-fold line skeleton curve models, and established corresponding restoring force models based on this. Wang [11] established the ideal trilinear skeleton curve model of RC shear wall members, and introduced the damage model into the hysteresis rule to establish the restoring force model of RC shear wall. Wei [12] and Zeng [13] established a four-line model of multi-ribbed composite wall considering stiffness degradation. Combined with statistical regression analysis, the corresponding restoring force model was established, and the influencing factors of the restoring force model were analyzed in depth. Song [14] established the restoring force model of shear wall with vertical slits by experimental fitting method. Ma [15] established a degraded three-line restoring force model of lattice composite shear walls, but did not consider the strength degradation; Lian et al. [16] presented a method for calculating characteristic parameters of restoring force model of composite shear wall. Kou al. [17] established a four-line restoring force model of fiber reinforced concrete shear wall considering stiffness degradation by taking cracking point, yield point, peak point and limit point as characteristic points. Li et al. [18, 19] carried out experimental research on reinforced concrete low and high shear walls and established corresponding restoring force models. The above research results of the restoring force model of shear wall members have great inspiration for this study, and will be a useful reference for the establishment of the restoring force model.

6.2 EHSW Skeleton Curve Model Under Monotonic Loading

Through observing the shape and trend of the skeleton curve of the EHSW specimen and combining it with the stiffness calculation results of the cracking stage, yield stage, and limit stage, it can be found that the stiffness of the specimen exhibits a clear turning point and degradation in the cracking, yield, and limit stages. Therefore, a four-line model, considering the cracking point, yield point, limit point, and model failure point (Fig. 6.1), is proposed to simulate the skeleton curve of the EHSW specimen. This four-line model enables a reasonable description of the load–displacement relationship throughout the monotonic loading process of the EHSW specimen until failure.

To establish the four broken line model, it is necessary to determine the load and displacement at the cracking point, yield point, limit point, and failure point.

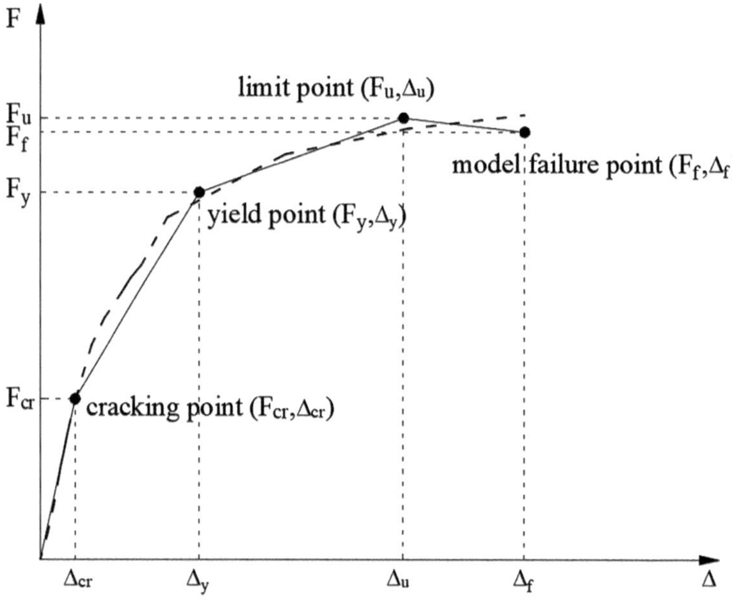

Fig. 6.1 Quadrilinear skeleton curve model

Table 6.1 Load and displacement of cracking point of EHSW specimen

Test piece	Cracking stage stiffness/(kN/mm)	Cracking point load/kN	Cracking point displacement/mm
EHSW1	44.59	301	6.75
EHSW2	44.59	252	5.65
EHSW3	44.59	262	5.88
EHSW4	44.59	268	6.01
EHSW5	44.59	245	5.50
EHSW6	44.59	293	6.57
EHSW7	44.59	304	6.82
EHSW8	36.91	225	6.10

6.2.1 Determination of Load and Displacement at Cracking Point

Using the cracking load calculation method for EHSW specimens in Sect. 5.1 and the stiffness calculation method for EHSW specimens in Sect. 5.4.1, the load and displacement at the cracking point can be directly determined. The displacement can be calculated from the load and stiffness at this stage, as shown in Eq. 5.22.

The calculation results for the cracking point load and displacement of each EHSW specimen are shown in Table 6.1.

6.2.2 Determination of Yield Point Load and Displacement

Based on the EHSW yield load calculation method in Sect. 5.2 and the EHSW yield stage stiffness calculation method in Sect. 5.4.2, the load and displacement of the yield point can be determined at this point. The displacement can be calculated from the load increment and stiffness of the stage, as well as the displacement of the cracking point.

The calculation results of the yield point load and displacement for each EHSW specimen are shown in Table 6.2.

6.2.3 Determination of Limit Point Load and Displacement

Based on the calculation methods for EHSW ultimate load in Sect. 5.3 and EHSW ultimate stage stiffness in Sect. 5.4.3, the load and displacement of the limit point can be determined. The displacement can be calculated using the load increment and

Table 6.2 EHSW specimen yield point load and displacement

Test piece	Yield stage stiffness/(kN/mm)	Yield point load/kN	Yield point displacement/mm
EHSW1	17.16	685.7	29.2
EHSW2	17.16	658.8	29.4
EHSW3	17.16	657.5	28.9
EHSW4	17.16	672.5	29.6
EHSW5	17.15	662.6	29.8
EHSW6	15.85	684.2	31.3
EHSW7	18.47	684.6	27.4
EHSW8	14.09	584.6	31.6

Table 6.3 The limit point load and displacement of EHSW specimen

Test piece	Ultimate stage stiffness/(kN/mm)	Limit point load/kN	Limit point displacement/mm
EHSW1	3.78	832.6	68.0
EHSW2	3.43	804.9	72.0
EHSW3	3.49	804.9	71.2
EHSW4	3.53	818.1	70.8
EHSW5	3.36	807.5	72.9
EHSW6	3.05	797.1	68.3
EHSW7	4.43	865.2	68.2
EHSW8	3.10	752.8	85.8

stiffness of the stage, as well as the displacement of the yield point, as shown in Eq. 5.28.

The calculation results for the limit point load and displacement of each EHSW specimen are shown in Table 6.3.

6.2.4 *Determination of Load and Displacement of Failure Point*

For the failure point load, based on the tested data, the limit load (*the load of the previous loading level in the failure stage*) can be determined by considering a proportional coefficient. The calculation method is shown in Eq. 6.1.

$$F_f' = \phi F_u' \tag{6.1}$$

In the formula:

F_f' failure load;
ϕ The ratio of the model failure load to the measured ultimate load under positive loading is detailed in Table 6.4;
F_u' The calculated value of the limit load based on the correction theory, and its calculation method is presented in Eq. 5.18.

The stiffness in the failure stage is still determined by the stiffness reduction (*reduction factor* φ) in the cracking stage. The specific calculation method is shown in Eq. 6.2. As can be seen from Table 6.4, the bearing capacity of EHSW4 and EHSW5 did not decrease in the failure stage, and the stiffness of the failure stage was positive, while the stiffness of other EHSW specimens was negative in this stage. According to the fitting analysis of the experimental data, the calculation results of the reduction coefficient φ are shown in Table 6.5.

$$K_f' = \varphi K_{cr}' \tag{6.2}$$

In the formula:

K_f' stiffness at failure stage;
φ The stiffness reduction coefficient in the failure stage is under the condition of positive loading;
K_{cr}' The theoretical calculated value of the stiffness correction during the cracking stage is presented in Formula .

In this stage, the displacement of the failure point can be calculated using load increments, stiffness, and limit point displacement, as shown in Eq. 6.3.

$$\Delta_f' = \Delta_u' + \frac{F_f' - F_u'}{K_f'} \tag{6.3}$$

Table 6.4 Measured results of EHSW specimens

Test piece	EHSW1	EHSW2	EHSW3	EHSW4	EHSW5	EHSW6	EHSW7	EHSW8
ϕ	0.998	0.963	0.963	1.020	1.055	0.997	0.988	0.995

Table 6.5 Theoretical results of EHSW specimens

Test piece	EHSW1	EHSW2	EHSW3	EHSW4	EHSW5	EHSW6	EHSW7	EHSW8
φ	– 0.002	– 0.031	– 0.028	0.016	0.048	– 0.002	– 0.009	– 0.014

Table 6.6 Load and displacement of failure point of EHSW specimen

Test piece	Stiffness in failure stage/(kN/mm)	Failure point load/kN	Displacement of failure point/mm
EHSW1	– 0.079	830.6	92.9
EHSW2	– 1.394	775.3	93.3
EHSW3	– 1.268	775.7	94.2
EHSW4	0.711	834.7	94.1
EHSW5	2.122	851.5	93.7
EHSW6	– 0.111	794.3	93.5
EHSW7	– 0.395	854.9	94.3
EHSW8	– 0.521	749.1	92.9

In the formula:

Δ_f' Displacement of failure point;
Δ_u' Limit point displacement, calculation method see formula ;
F_f' The calculation value of failure load is shown in Formula 6.1;
F_u' Limit load correction theoretical calculation value, the calculation method is shown in Eq. ;
K_f' Stiffness calculation value in failure stage, calculation method is shown in 6.2.

The calculation results of load and displacement at the failure point of each EHSW specimen are shown in Table 6.6.

6.2.5 Four—Fold Skeleton Curve Model

The skeleton curve models of each EHSW specimen and the hysteresis and skeleton curves of the test are depicted in Fig. 6.2 based on the calculation results of Sects. 6.2.1–6.2.4. It can be inferred from the figure that the four-fold line model aligns well with the test curve, capturing the principal stages of stress and deformation in the EHSW specimen under horizontal loading.

6.3 EHSW Restoring Force Model Under Cyclic Loading

6.3.1 Hysteresis Rules

The establishment of the restoring force model for the component under reciprocating load should first determine its hysteresis rules, including the unloading stiffness of the component and the forward and reverse reloading routes.

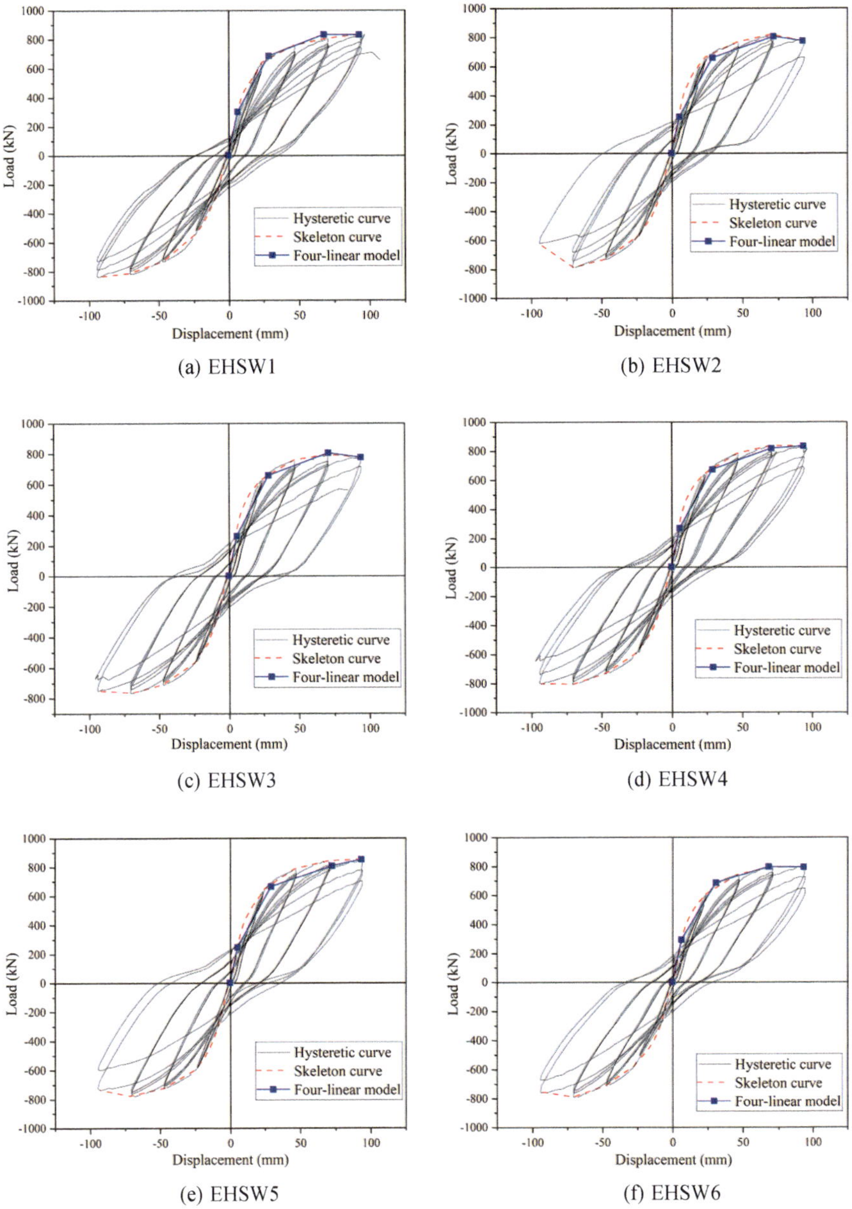

Fig. 6.2 Calculation results of EHSW specimen skeleton curve four-line model

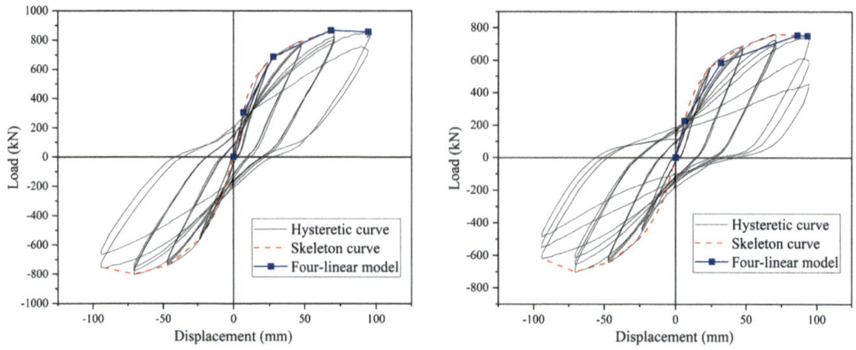

(g) EHSW7 (h) EHSW8

Fig. 6.2 (continued)

By examining the test hysteresis curve of an EHSW specimen, it is observed that:

① When the load does not exceed the yield load, the hysteresis loop unloading curve is parallel to the initial loading (*cracking stage*) curve, and the unloading stiffness does not change significantly. Therefore, the unloading stiffness of the hysteresis curve until the yield point stage can be taken as the loading stiffness in the cracking stage, namely $K_{y,u} = K'_{cr}$

② When the load is further increased and exceeds the yield load, the stiffness of the hysteresis loop unloading curve begins to degenerate. The unloading stiffness of each EHSW specimen at the vertex of each hysteresis loop $K_{u,i}$ (i represents the loading level) is calculated. Based on existing experience, and numerical fitting methods, the fitting formula $K_{u,i}$ is established. The specific formula is 6.4, and the correlation coefficient between the fitting formula and the experimental value is $R^2 > 0.998$.

$$\frac{K_{u,i}}{K'_{cr}} = 0.57461\left(\frac{\Delta_i}{\Delta_y}\right)^{-0.48228} \tag{6.4}$$

In the formula:

$K_{u,i}$ stage *i* loading displacement unloading stiffness;
K'_{cr} The elastic stage stiffness, namely the cracking stage stiffness, is calculated according to Formula ;
Δ_i Level *i* loading displacement;
Δ_y Yield point displacement.

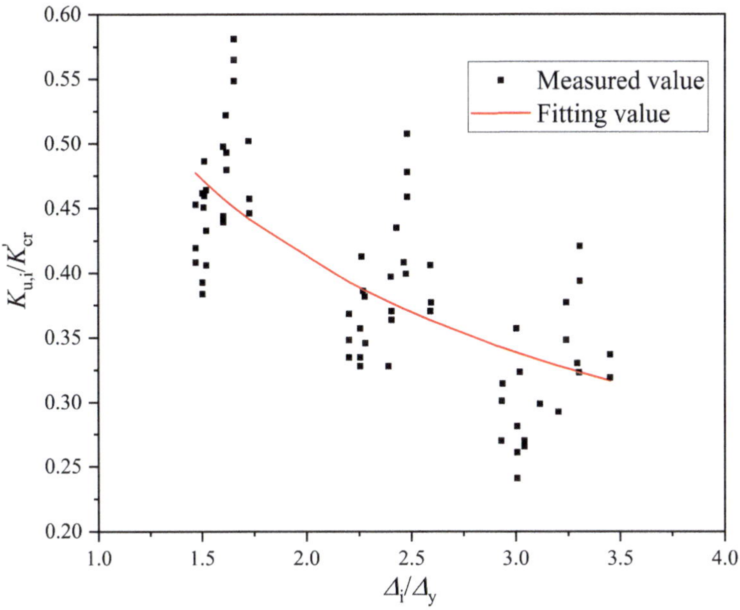

Fig. 6.3 Fitting calculation results of unloading stiffness after yield point of EHSW specimen

By observing the hysteresis curve of EHSW specimen, it can be found that positive, reverse, and repeated loading all point to a fixed point on the skeleton curve, which belongs to the fixed point pointing type (Fig. 3.20). Through statistical analysis, the ordinate value of this point, i.e., the load value, exhibits a good linear correlation with the ordinate value of the cracking point on the skeleton curve, with F'_{cr} as the reference point (Fig. 6.3).

6.3.2 Hysteretic Route

Following the hysteresis rule and skeleton curve model, the walking route of the EHSW specimen during loading, unloading, and reloading is plotted, as well as the hysteresis loop of the EHSW specimen, as shown in Fig. 6.4.

(1) Walking route parameter description

① $\pm F_{cr}$, Four broken line skeleton curve model cracking point load, calculation method is shown in Formula 5.5;
② $\pm F_y$, four broken line skeleton curve model yield point load, calculation method is detailed in Formula 5.17;
③ $\pm F_u$, four broken line skeleton curve model limit point load, calculation method is shown in Formula 5.18;

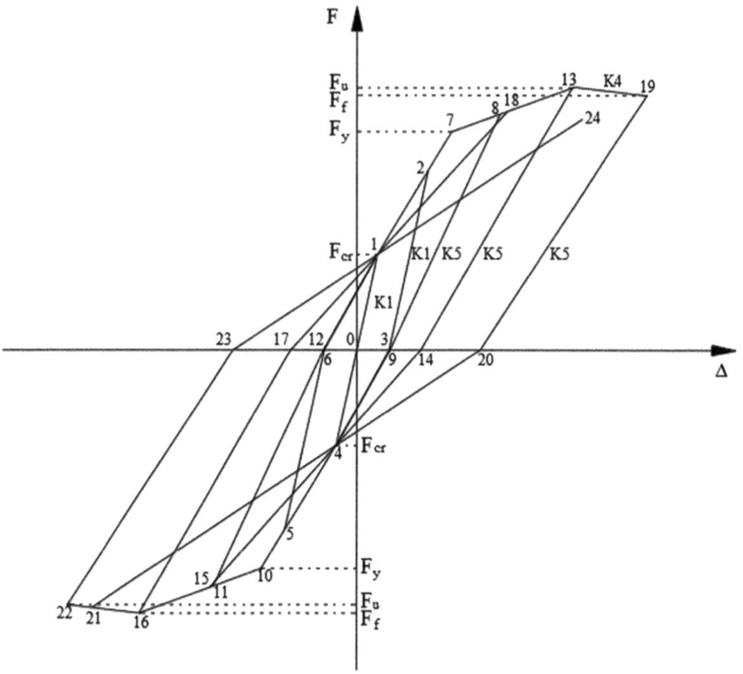

Fig. 6.4 Hysteretic route of EHSW specimen

④ $\pm F_f$, four broken line skeleton curve model failure point load, calculation method is shown in Formula 6.1;

⑤ $K1$, elastic stiffness is cracking stage stiffness, or loading to yield point before unloading stiffness, calculation method is shown in 5.21;

⑥ $K2$, The stiffness from cracking to pre-yield, that is, the stiffness at the yield stage, is calculated by Eq. 5.24.

⑦ $K3$, yield to the load reaches the limit stiffness, namely the limit stage stiffness, calculation method is shown in Eq. 5.27;

⑧ $K4$, the load reaches the limit value to the failure value stiffness, namely the failure stage stiffness, the calculation method is shown in Formula 6.2;

⑨ $K5$ After loading to the yield point, the unloading stiffness is calculated by formula 6.4.

(2) Walking route description

The loading, unloading, and repeated loading and unloading curves of the restoring force model should be confined within the four-fold skeleton curve. Based on the loading level, the walking route of the restoring force curve is detailed according to the following four conditions:

1. before cracking, the component is in the elastic stage, the loading and unloading curves are straight lines, and the loading, unloading, and repeated loading paths are 0 → 1 → 0 → 4 → 0;
2. loading exceeds the cracking point but does not reach yield, and the component enters the elastoplastic stage. The unloading stiffness of this stage is the same as that of the elastic stage. In the reverse reloading process, the curve must pass through the cracking point, and the walking route is 0 → 1 → 2 → 3 → 4 → 5 → 6 → 1 → ...;
3. when loading exceeds yield but does not reach the limit, the component's plasticity further develops, and the unloading stiffness degrades at this stage. In the reverse reloading process, the curve must pass through the cracking point and may intersect the skeleton curve in advance before reaching the loading load or displacement. The walking route is 0 → 1 → 2 → 7 → 8 → 9 → 4 → 10 → 11 → 12 → 1 → 7 → 13 → ...;
4. loading exceeds the limit but does not reach failure, and the stage of unloading stiffness degradation, take the same, the reverse re-loading process, the curve must pass through the cracking point, and may not reach the loading load or displacement before, and the skeleton curve intersected in advance, the route is 0 → 1 → 7 → 13 → 19 → 20 → 4 → 21 → 22 → 23 → 1 → 24 → ...

6.3.3 EHSW Restoring Force Model

Establishing a restoring force model according to skeleton curve and hysteresis rules, and a specific walking route termed hysteresis route, enables comparison with test hysteresis curves. Figure 6.5 illustrates the comparison between the restoring force model of each EHSW specimen and the corresponding test hysteresis curve.

Figure 6.5 shows good agreement between the established restoring force model and the test hysteresis curve, reflecting the load–displacement relationship characteristics of the EHSW specimen throughout the testing process of loading, unloading, and reloading under low cyclic loading. The model also exhibits good fit to the unloading stiffness degradation.

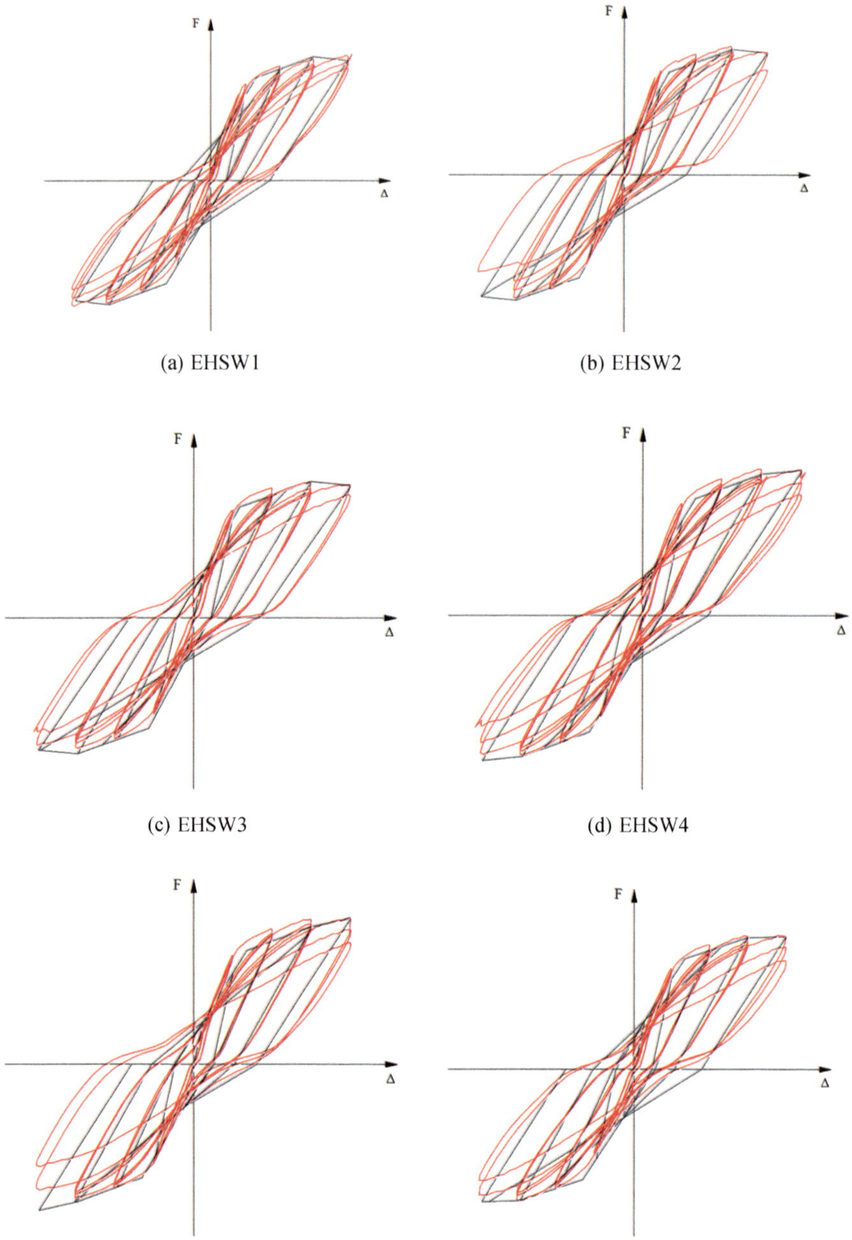

(a) EHSW1 (b) EHSW2

(c) EHSW3 (d) EHSW4

(e) EHSW5 (f) EHSW6

Fig. 6.5 Restoring force model of EHSW specimen (the imaginary line is the test hysteresis curve, and the real line is the restoring force model)

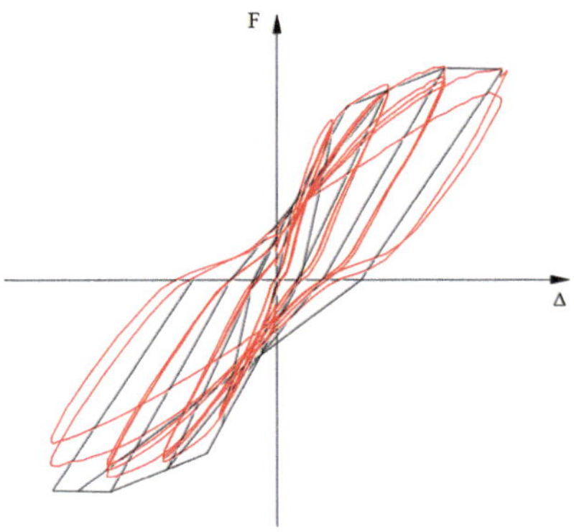

(g) EHSW7

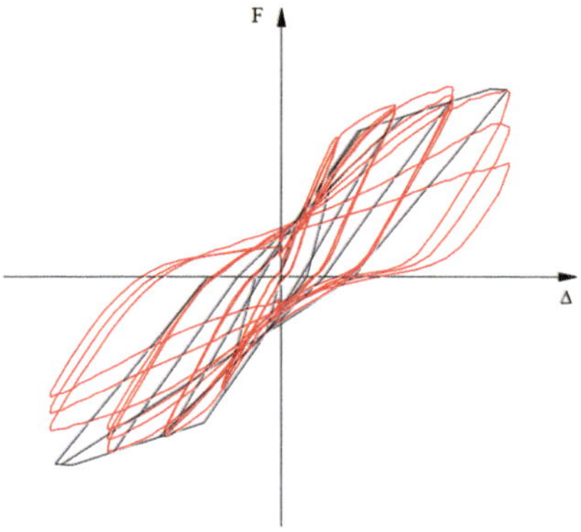

(h) EHSW8

Fig. 6.5 (continued)

References

1. Lv XL, Jin GF, Wu XH. Nonlinear finite element theory and application of reinforced concrete structures. Shanghai: Tongji University Press; 1997.
2. Guo ZH, Shi XD. Principle and analysis of reinforced concrete. Beijing: Tsinghua University Press; 2003.
3. Jiang HJ, Lv XL. Analysis of shear wall structures using a wall element model. Earthq Eng Eng Vib. 1998;18(3):40–48.
4. Qian JR, Xu FJ. Displacement-based deformation capacity design method for reinforced concrete shear walls. J Tsinghua Univ Nat Sci Ed. 2007;47(3):1–4.
5. Priestley MJN. Aspect of drift and ductility capacity of rectangular cantilever structural walls. Bull N Z Soc Earthq Eng. 1998;31(2):73–85.
6. Zhang GJ, Lv XL and Liu BQ. Research on restoring force model of frame column when axial compression ratio exceeds the limit. J Build Struct. 2006;27(1): 90–98.
7. Li L. Experimental and theoretical study on seismic behavior of high strength concrete-steel composite shear wall. Beijing: Beijing University of Technology; 2008.
8. Liu HL. Study on seismic behavior of double-layer steel plate infilled concrete composite shear wall with binding rod. Guangzhou: South China University of Technology; 2013.
9. Li XL. Mechanical model and experimental study on short-pier shear wall structure of high-rise building. Xi'an: Xi'an University of Architecture and Technology; 2011.
10. Bai L. Study on seismic performance and performance design theory of steel reinforced high performance concrete shear wall. Xi'an: Xi'an University of Architecture and Technology; 2009.
11. Wang K. Experimental study on restoring force model of reinforced concrete shear wall based on damage. Xi'an: Xi'an University of Architecture and Technology; 2011.
12. Wei X. Study on restoring force model of multi-ribbed composite wall structure. Xi'an: Xi'an University of Architecture and Technology; 2008.
13. Zeng H. Research on restoring force model and influencing factors of multi-ribbed composite wall structure. Xi'an: Xi'an University of Architecture and Technology; 2011.
14. Song WS. Nonlinear finite element analysis and restoring force model of shear wall slabs with vertical slits. Tianjin: Tianjin University; 2009.
15. Ma F. Experimental study on the influence of axial compression ratio on the seismic performance of lattice composite shear walls. Tianjin: Tianjin University; 2000.

16. Lian X, Ye XG, Jiang Q. Calculation method of characteristic parameters of restoring force model of composite slab shear wall. J Huaqiao Univ (Nat Sci Ed). 2010;31(1):88–94.
17. Kou JL, Liang XW, Deng MK. Model test and theoretical study on restoring force of fiber reinforced concrete shear wall. 2013;46(10): 58–70.
18. Li B, Li HN, Cao JD. Quasi-static test of reinforced concrete high shear wall. J Shenyang Jianzhu Univ (Nat Sci Ed). 2009;25(2):230–234.
19. Li B, Li HN, Cao JD. Quasi-static test and hysteretic model of reinforced concrete low shear wall. J Shenyang Jianzhu Univ (Nat Sci Ed). 2010;26(5):869–874.

Chapter 7
Structural Improvement of Emulative Hybrid Assembled Precast Concrete Shear Wall Based on OpenSees

The test of emulative hybrid assembled precast concrete shear walls (EHSW) and analysis of ABAQUS parameters have highlighted the benefits of incorporating prestressed connection and grout anchored overlap technologies to enhance component seismic performance. However, the limitations of the test specimens and ABAQUS analysis have constrained the full exploration of seismic performance indicators such as hysteresis behavior, ductility, and energy dissipation capacity of the EHSW. To overcome these constraints, the OpenSees analysis software is employed to delve into the genuine seismic performance of EHSW. This approach is rooted in the concept of assembled monolithic precast concrete shear wall and pertinent structural enhancements, moving beyond the conventional notion of "equivalent cast-in-place". Thus, a pursuit of more efficient, rational, and dependable EHSW structures is embarked upon.

7.1 Establishment of OpenSees Model

7.1.1 Analysis Software Selection

OpenSees (*Open System for Earthquake Engineering Simulation*) is a widely used finite element analysis software and earthquake engineering simulation platform in civil engineering academia. The software was developed by the Pacific Earthquake Engineering Research Center (*PEER*) under the National Science Foundation (*NSF*), led by the University of California, Berkeley, and nearly 10 leading American universities [1].

As a new generation of finite element calculation software, the program is written using object-oriented advanced program architecture. Users can carry out secondary development of element models, material constitutive models, iterative algorithms, and post-processing forms. Its script uses Tcl/Tk language, Tcl/Tk is a

Z. Zhu et al., *Research on Emulative Hybrid Assembled Precast Concrete Shear Walls*,
https://doi.org/10.1007/978-981-96-9578-2_7

fully programmable scripting language, which is used to define the model, solve the process, and post-process. It can solve not only simple problems, but also model and study parameters of large and complex systems. Users can use Visual C++ for secondary development, and can also use dynamic API to call new modules (*DLL file form*) and functions when the program runs. Analysis software selection.

OpenSees has evolved into one of the most influential open research platforms in seismic engineering with its powerful nonlinear numerical simulation capabilities, rich material and unit libraries, a variety of efficient algorithms, open program architecture, and advanced concepts for continuous integration of the latest research results. It is widely used in many fields such as nonlinear simulation and seismic analysis of structures, numerical calculation and seismic response simulation of geotechnical engineering, structural sensitivity and reliability analysis, structural collapse analysis, friction collision analysis, and hybrid test simulation.

In order to make up for the limitations of ABAQUS analysis results and truly reproduce the seismic performance of the improved EHSW, the seismic performance indexes such as hysteretic performance, ductility, and energy dissipation capacity are analyzed. The fiber element model is established based on the OpenSees platform, and the horizontal low cyclic loading test of the simulation test is carried out, and the seismic performance is evaluated by analyzing the hysteretic curve results.

7.1.2 Element Type Selection

In OpenSees' entire analysis program, the element object is the most critical component. The program provides users with a variety of macro element types, including three common beam-column elements for shear wall component simulation (stiffness-based beam-column element, flexibility-based beam-column element, and flexibility-based plastic hinge beam-column element), a truss element for steel bar simulation, and a zero-length section element for specialized applications.

(1) Beam-column element based on stiffness method

The nonlinear element (Disp-beam column element) based on the finite element stiffness method employs the displacement interpolation function as its interpolation function. The solution approach involves dividing the element into several integral sections, using cubic Hermite to construct the tangential displacement field, and deriving the interpolation function to obtain the corresponding section deformation at each section, as shown in Formulas 7.1 and 7.2.

$$d(x) = \begin{Bmatrix} \ddot{v}(x) \\ \dot{u}(x) \end{Bmatrix} = \overline{\alpha}(x)\overline{q} \tag{7.1}$$

$$\overline{\alpha}(x) = \begin{bmatrix} \ddot{\phi}_1(x) & \ddot{\phi}_2(x) & 0 & \ddot{\phi}_3(x) & \ddot{\phi}_4(x) & 0 \\ 0 & 0 & \dot{\varphi}_1(x) & 0 & 0 & \dot{\varphi}_2(x) \end{bmatrix} \tag{7.2}$$

Adopting the fiber model in the section enables the calculation of the strain and stress for each fiber, providing insights into the section's resistance and stiffness. The element's stiffness matrix can be determined by integrating the section's stiffness over its length, as shown in Eq. 7.3. Similarly, the section's resistance vector is integrated along the length to obtain the element's resistance vector, as shown in Eq. 7.4.

$$\overline{K} = \int_0^L \overline{a}^T(x)k(x)\overline{a}(x)dx \tag{7.3}$$

$$\overline{Q}_R = \int_0^L \overline{a}^T(x)D_R(x)dx \tag{7.4}$$

Since axial strain is the first derivative of axial displacement and curvature the second of tangential displacement, constructed displacement fields can lead to issues of constant axial strain, linear curvature and curvature discontinuity. Generally, errors from displacement fields can be reduced by unit subdivision, but this increases structure degrees of freedom and calculation cost. The main disadvantage of stiffness-method elements is that cubic Hermite interpolation poorly describes curvature after yield, and there is no element-level iteration, so convergence is slow. To reduce Hermite-function errors, multi-subdivision modeling is used, which yields good results.

(2) Beam-column element based on flexibility method

The nonlinear beam-column element based on the finite element flexibility method differs from the stiffness-based method in that the model uses force linear interpolation rather than displacement interpolation at the integration point. The interpolation function used is shown in Eq. 7.5.

$$b(x) = \begin{bmatrix} x(L-1) & xL & 0 \\ 0 & 0 & 1 \end{bmatrix} \tag{7.5}$$

The element force is linearly converted to cross-sectional force by an interpolation function. This cross-sectional force is then converted to deformation via the cross-sectional flexibility matrix obtained from the previous iteration. When the cross-section adopts a fiber model, the strain and corresponding stress of each fiber in the cross-section are obtained, enabling calculation of the cross-sectional resistance and tangent stiffness. Integrating the flexibility matrix of the section along its length yields the flexibility matrix of the element, as shown in Formula 7.6.

$$\overline{F} = \textstyle\int_0^L b^T(x)f(x)b(x)dx \tag{7.6}$$

When the cross-section resistance and the external force of the cross-section do not meet the tolerance requirements, the unbalanced force of the cross-section is transformed into residual deformation of the cross-section. This is then transformed into the deformation increment of the next iteration of the element through Gaussian integration. When the cross-section resistance equals the cross-section force, the cross-section is balanced, which is called internal iteration of the element. Due to the internal iterative calculation of the element, the convergence speed of the overall structure calculation is greatly improved. Nonlinear beam-column elements based on stiffness method and flexibility method mostly adopt Gauss–Lobatto integral method. The position and weight of integral points are shown in Fig. 7.1.

Compared to the beam-column element of the stiffness method, this element is not limited by linear curvature distribution. When simulating curved beam-column members, it can achieve good results and fast convergence, without the need for further member subdivision.

(3) Plastic hinged beam-column element based on flexibility method

A large number of experiments show that the plastic hinge of beam-column mostly occurs at the end, with the middle part remaining essentially elastic. Consequently, a beam element incorporating hinges is proposed using flexibility method. Similar to the calculation process for beam-column elements using flexibility method, internal iteration for the element is required. However, the middle integral section of the plastic hinge element adopts elastic constitutive, and the internal iteration only needs

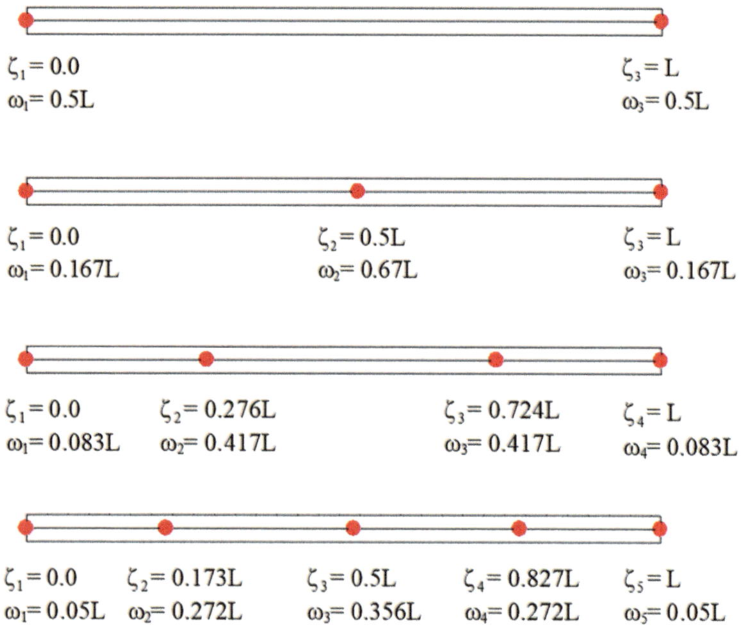

Fig. 7.1 Gauss–Lobatto integral points and weights [1]

to be conducted in the plastic hinge area, facilitating convergence and improving solution speed.

(4) Truss element

The Corotational truss element can only transmit axial forces, and compared to the earlier bar elements in OpenSees, the advantage of the truss element is that it can consider the effect of geometric nonlinearity on the deformation of the beam element, so the calculation is more accurate.

(5) Zero-length section element

The stress and deformation state of the zero-length section element is determined by the state of its fiber section. Since the length of the element is a unit length, the rotation deformation of the element is numerically equal to the curvature response of its fiber section, and the internal force of the section is integrated to obtain the internal force of the element. The element is often used to simulate the effectiveness of the interaction between two structural components, such as beam-column and column-foundation, and the mechanical properties between joints at the bottom of precast shear walls.

Combining the advantages and disadvantages of the above several nonlinear beam-column elements, this book uses a beam-column element based on the flexibility method to simulate shear wall members and a truss element to simulate prestressed reinforcement. Additionally, a zero-length section element is added at the bottom of the EHSW wall to consider the influence of the mechanical properties of the joint on the overall component.

7.1.3 Material Constitutive Model

(1) Constitutive model of concrete

OpenSees employs the Concrete01 uniaxial material, based on the modified Kent-Park concrete constitutive model, which ignores the tensile capacity of concrete. It considers the peak stress of the compressive end of concrete, peak strain, and softening curvature of the descending section to reflect the constraints of stirrups. This model can also account for the residual strength of concrete. In general, it is a simple and accurate concrete model. The stress–strain relationship during unloading is determined using the Karsan-Jirsa unloading rule. The damage of concrete is considered by the attenuation of the slope of the straight line in the unloading section, with the notable feature that unloading and reloading are along the same line. The compression curve is divided into three sections: rising, falling, and platform, as shown in Fig. 7.2, and the skeleton expression is shown in Eqs. 7.7–7.12.

$$\text{climb element } (\varepsilon \leq \varepsilon_0) \quad \sigma = Kf_{co}\left[2\left(\frac{\varepsilon}{\varepsilon_0}\right) - \left(\frac{\varepsilon}{\varepsilon_0}\right)^2\right] \tag{7.7}$$

$$\text{sloping portion } (\varepsilon_0 < \varepsilon \le \varepsilon_{cu}) \quad \sigma = Kf_{co}[1 - Z(\varepsilon - \varepsilon_0)] \tag{7.8}$$

$$\text{platform segment } (\varepsilon > \varepsilon_{cu}) \quad \overline{F} = \int_0^L b^T(x)f(x)b(x)dx \tag{7.9}$$

$$\varepsilon_0 = 0.002K \tag{7.10}$$

$$K = 1 + \rho_v f_{yh}/f_{co} \tag{7.11}$$

$$Z = \frac{0.5}{\frac{3+0.29f_{co}}{145f_{co}-1000} + 0.75\rho_v\sqrt{\frac{h'}{s_h}} - 0.002K} \tag{7.12}$$

In the formula:

ε_0 is the compressive strain corresponding to the peak stress of concrete;
ε_{cu} is the corresponding compressive strain when the concrete stress decreases to 20% of the peak stress;
K the improvement coefficient of stirrup to concrete strength;
Z he strain softening slope coefficient;
f_{co} compressive strength of concrete cylinder;
f_{yh} the yield strength of stirrup (MPa);
ρ_v volumetric stirrup ratio for stirrups;
h' the limb spacing (mm) of the stirrup;
s_h The stirrup spacing (mm).

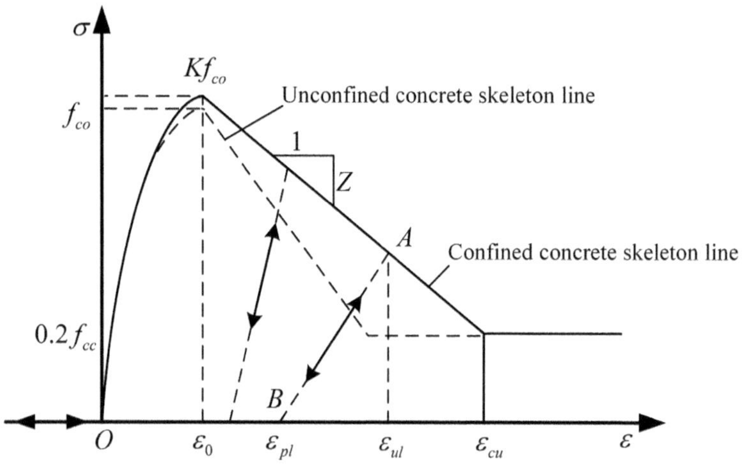

Fig. 7.2 Kent-park concrete constitutive [2]

Scott et al. [3] conservatively took the concrete strain value when the first confined stirrup fractured as the ultimate compressive strain value of the core concrete, took the strain value when the cover concrete fell off and failed as 0.004, and determined the ultimate compressive strain of confined concrete according to Formula 7.13.

$$\varepsilon_{cu} = 0.004 + 0.9\rho_v(f_{yh}/300) \tag{7.13}$$

In the formula:

ρ_v volume stirrup ratio of confined stirrup;
f_{yh} Yield stress of confined stirrups.

The compressive strength of concrete in OpenSees is the standard value of cylindrical compressive strength. For concrete below C60, the standard value of cylindrical compressive strength is 0.79 times the standard value of cubic compressive strength. The material parameters of confined and unconfined concrete in the core area are calculated using the modified Kent-Park model. The characteristic values of the concrete constitutive relation used in this analysis model are shown in Table 7.1.

2) Steel constitutive model

Both ordinary steel bar and prestressed steel bar adopt the Steel02 uniaxial material in OpenSees, which is based on the Pinto steel bar constitutive model. This model exhibits good numerical stability due to the explicit function expression of strain, and it is consistent with the results of repeated loading tests, thus better capturing the Bauschinger effect. The constitutive curve is shown in Fig. 7.4.

The curve gradually changes from the initial asymptote of the initial slope E_0 (the elastic modulus of the steel bar) to the yield asymptote of the slope E_1, $E_1 = bE_0$, b is the hardening coefficient of the steel bar, taken as 0.01.

The determined value of the stress–strain relationship expression depends on the reverse loading point $(\varepsilon_r, \sigma_r)$ and parameter R under the current stress state. The expression of the transition curve is shown in 7.14.

$$\sigma^* = b\varepsilon^* + \frac{(1-b)\varepsilon^*}{(1+\varepsilon^{*R})^{1/R}} \tag{7.14}$$

Table 7.1 Characteristic parameters of constitutive relation of OpenSees concrete

Concrete type	Peak strength/MPa	Peak strain	Ultimate strength/ MPa	Ultimate strain
Unconstrained zone concrete	27.97	0.002	10.87	0.004
Concrete in restraint area	41.50	0.0029	7.38	0.0445

The calculated constitutive model curve of concrete is shown in Fig. 7.3

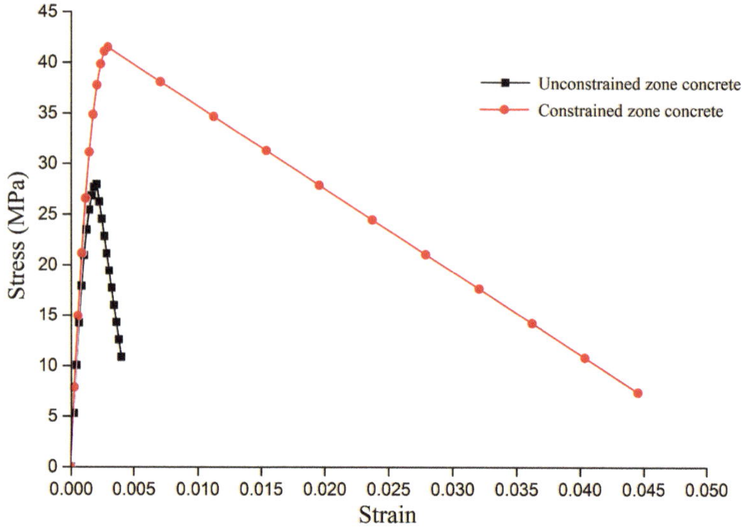

Fig. 7.3 OpenSees model concrete constitutive

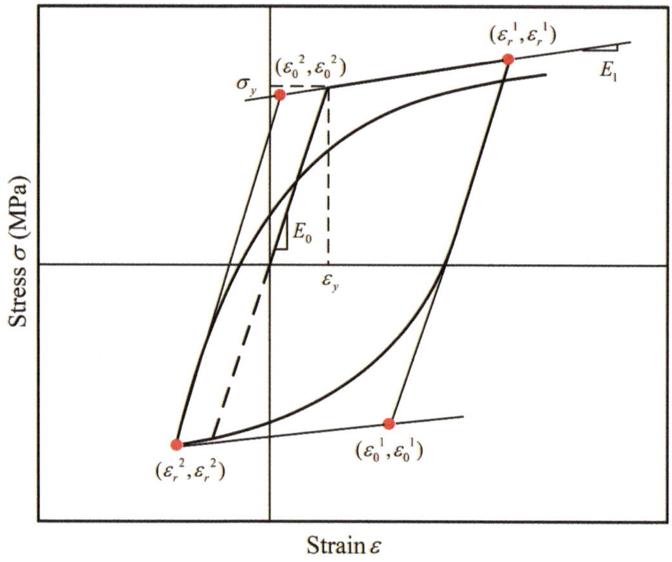

Fig. 7.4 Menegotto-Pinto steel constitutive [2]

where σ^* and ε^* are the normalized stress and strain, respectively, and the parameter R is the curvature coefficient of the transition curve. The Bauschinger effect of steel bars can be adjusted by this parameter, which is calculated according to Eqs. 7.15–7.17.

$$\sigma^* = \frac{\sigma - \sigma_r}{\sigma_0 - \sigma_r} \tag{7.15}$$

$$\varepsilon^* = \frac{\varepsilon - \varepsilon_r}{\varepsilon_0 - \varepsilon_r} \tag{7.16}$$

$$R = R_0 - \frac{a_1 \xi}{a_2 + \xi} \tag{7.17}$$

In the above formula, R_0 is the curvature coefficient of curvature at initial loading, a_1 and a_2 are the degradation coefficients of curvature at cyclic loading, and ξ is the maximum strain parameter in strain history, which is calculated according to Formula 7.18.

$$\xi = \left| \frac{\varepsilon_m - \varepsilon_0}{\varepsilon_y} \right| \tag{7.18}$$

In the above formula, ε_m represents the maximum strain within the strain history, while ε_y stands for the yield strain of the steel bar.

R_0, a_1, and a_2 refer to the recommended values provided by Menegotto and Pinto, which are 20, 18.5, and 0.15, respectively.

7.1.4 Model Establishment

Based on the actual geometric information of the specimen and the measured index value of the material properties in the test, corresponding numerical analysis models are established for the cast-in-place member and the EHSW member respectively. The model loading is displacement loading, and the loading system is consistent with the test process.

(1) Cast-in-place model

As shown in Fig. 7.5, the cast-in-place model adopts the beam-column element based on the flexibility method to simulate the shear wall. The wall member is divided into three elements along the vertical direction, and the number of element integration points is selected as 5. The beam-column element adopts the fiber section, which is endowed with the properties of unrestrained concrete, restrained concrete and steel bar respectively. The concrete and steel bar constitutive materials are concrete01 and Steel02 uniaxial constitutive materials respectively.

(2) EHSW model

As shown in Fig. 7.6, the EHSW model wall body is simulated using the same method as the cast-in-place component, and the prestressed reinforcement is simulated using the Corotational truss element, considering geometric nonlinearity. Since

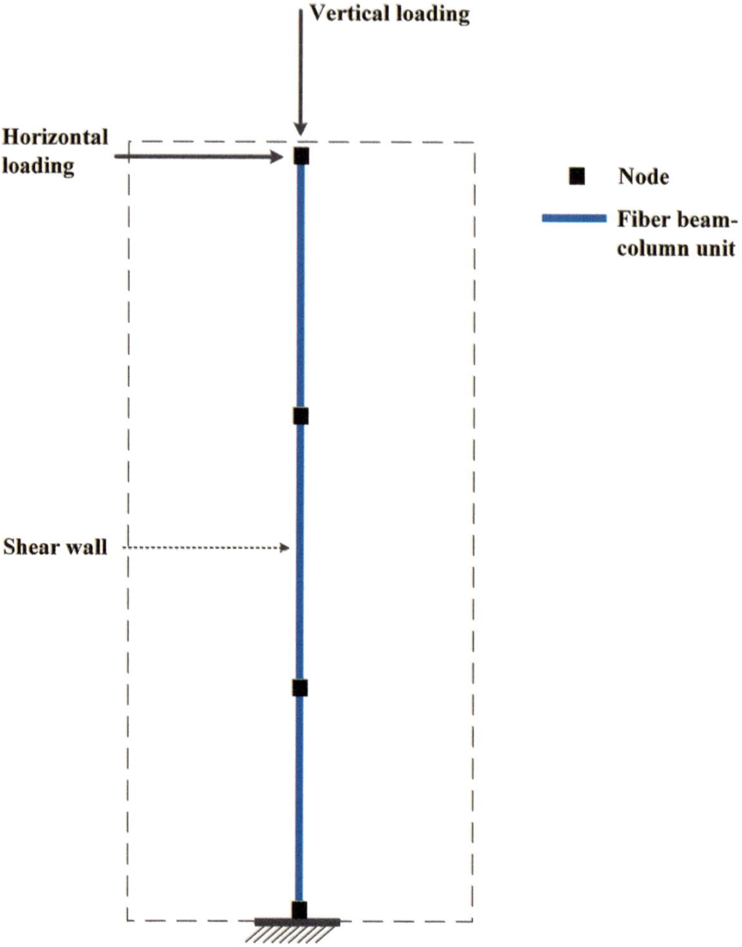

Fig. 7.5 Cast-in-place specimen OpenSees model

the prestressed reinforcement can only be pulled and not compressed, its constitutive uses Steel02 and ElasticPPGap materials in series, with the prestress applied by setting the initial prestress. To maintain consistent line shape with prestressed pipe during wall deformation, the unit division corresponds to the wall unit division. The bottom and top joints of unbonded prestressed steel bar are fully constrained with corresponding wall joints, with other joints coupled to the two horizontal translational degrees of freedom of the corresponding wall joints, and the vertical degree of freedom relaxed. Friction between the reinforcement and pipe is neglected. To simulate the unbonded segment of bottom grout-anchored rebar, the grout-anchored rebar in this segment is isolated from the fiber section. The truss element is used to simulate the actual position point, with the truss element node and the corresponding wall node kinematically constrained. At the wall bottom, the Concrete01 constitutive

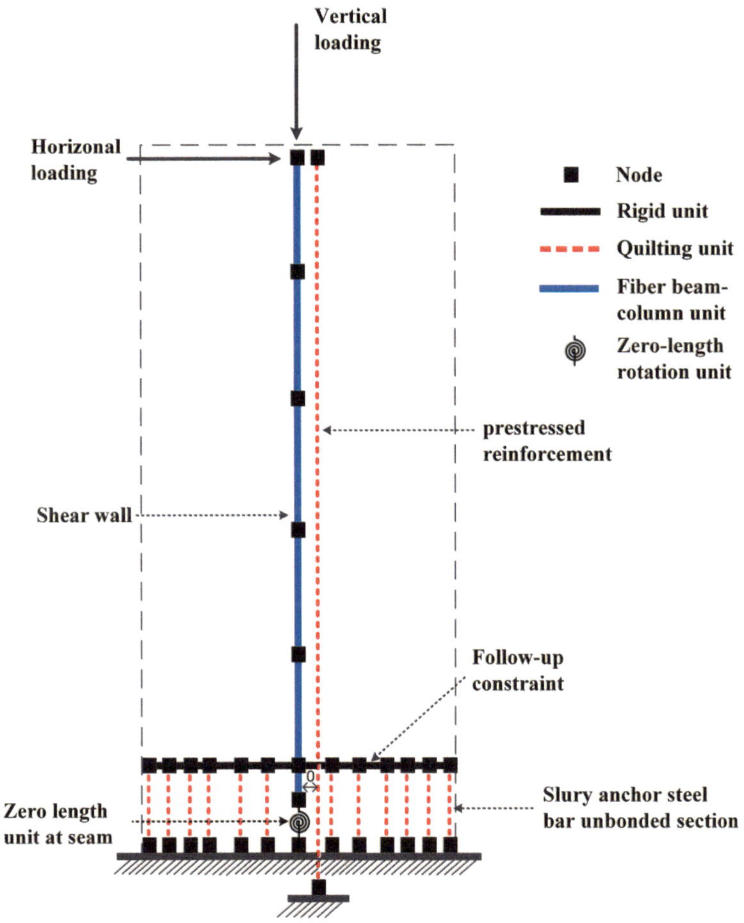

Fig. 7.6 OpenSees model of EHSW specimen

model, without considering concrete tensile strength, is used to simulate the EHSW wall bottom response.

7.2 OpenSees Model Validation

To ensure the accuracy of the finite element model and the reproducibility of test results, finite element analysis results are compared with test results, including the load–displacement hysteresis curve and the load–displacement skeleton curve.

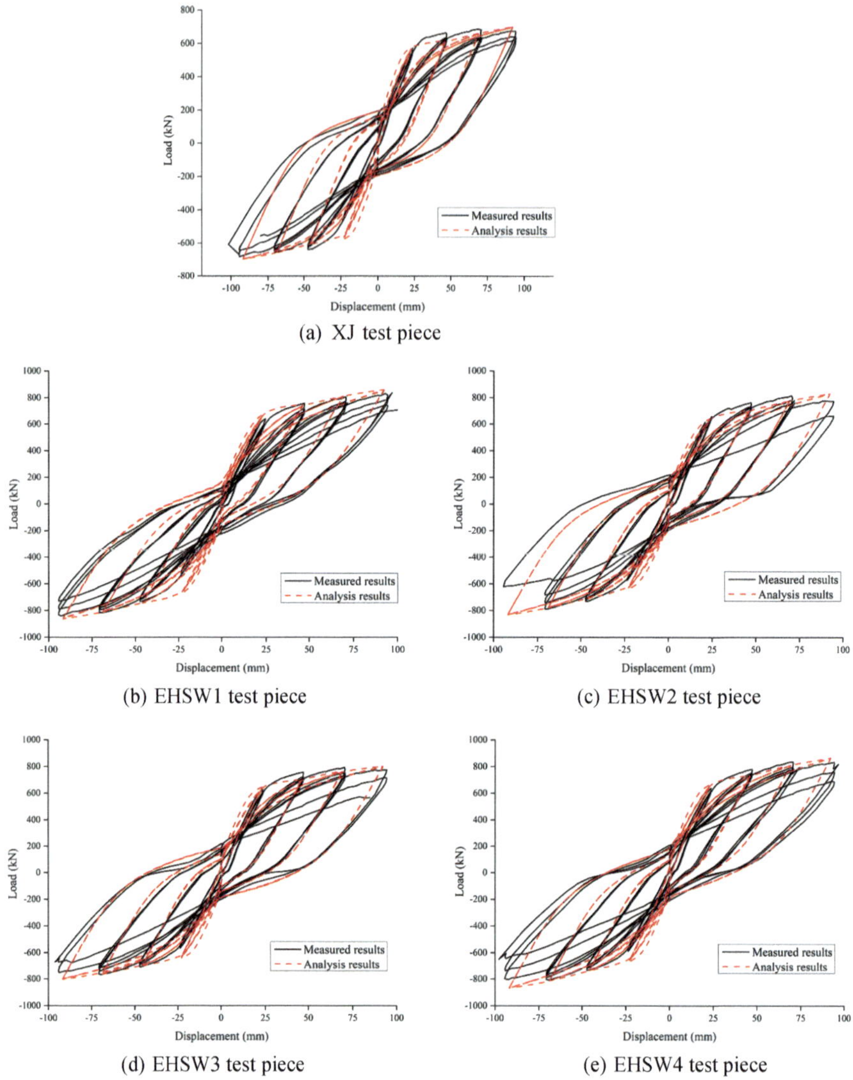

Fig. 7.7 Hysteresis curve comparison

The finite element simulation hysteresis curve and skeleton curve of each test component is compared with the test curve, as shown in Figs. 7.7 and 7.8. It can be seen from the figures that the pinching phenomenon and unloading process of the hysteresis curve of each test component are well simulated, and the load values of the numerical model in each loading stage are in good agreement with the measured

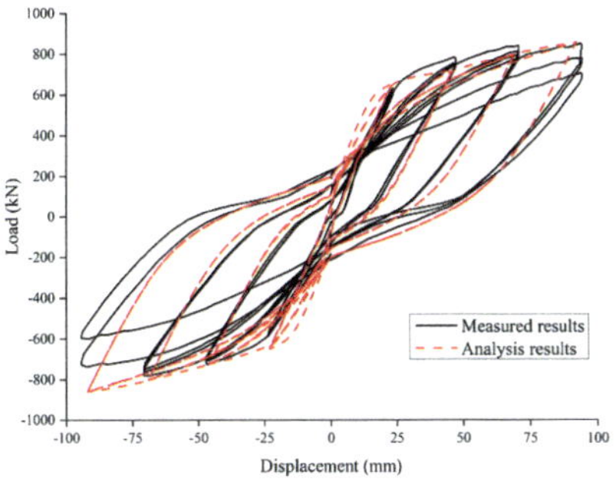

(f) EHSW5 test piece

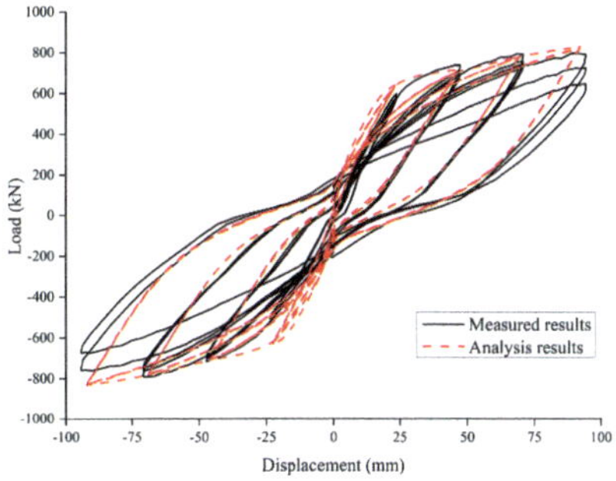

(g) EHSW6 test piece

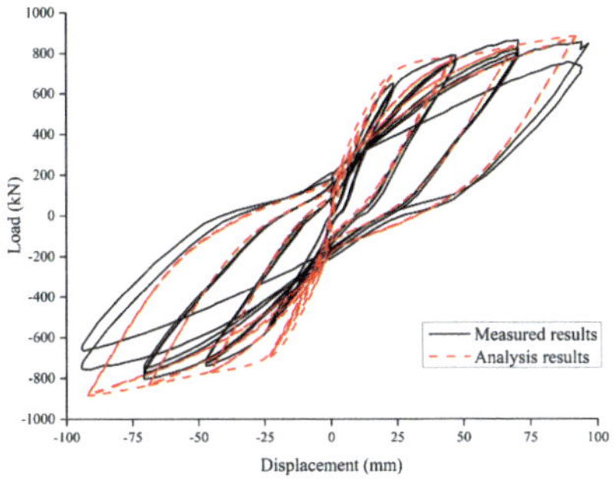

(h) EHSW7 test piece

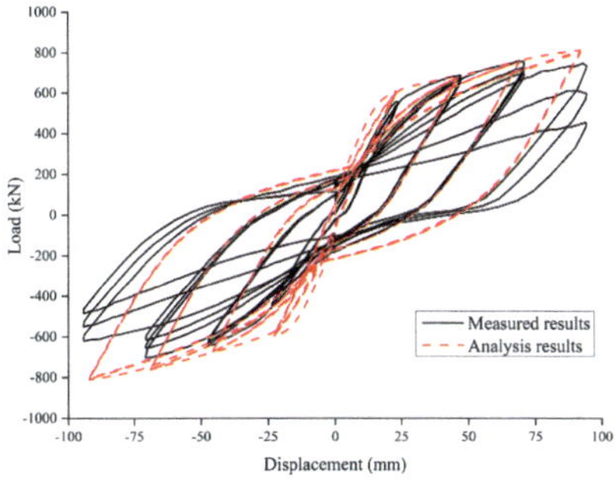

(i) EHSW8 test piece

Fig. 7.7 (continued)

values. Due to the beam-column element adopts the fiber section, the influence of shear deformation cannot be considered, so the simulation curve does not appear the phenomenon of falling failure in the final stage, and the unloading section of the simulated hysteresis curve is slightly faster than the test. At the same time, the horizontal bearing capacity at the later stage of loading is slightly larger than the measured value of the test due to the failure to consider the decline of steel bar stress from yield to fracture, but the overall trend meets the requirements of the test for numerical simulation. The initial loading and unloading stiffness in the numerical model is greater than the measured value of the test. One of the reasons for this phenomenon may be that there are original defects in the pouring and installation of the components and there are gaps and force deformation in the loading tooling, and the relevant effects are not simulated in the numerical simulation. The second is that the finite element analysis model does not deduct the prestress loss of the prestressed steel beam, but maintains the initial value in the test in an ideal condition. Therefore, the initial stiffness of the finite element model is large, but with the increase of the loading level, the deformation of the specimen itself gradually increases, and

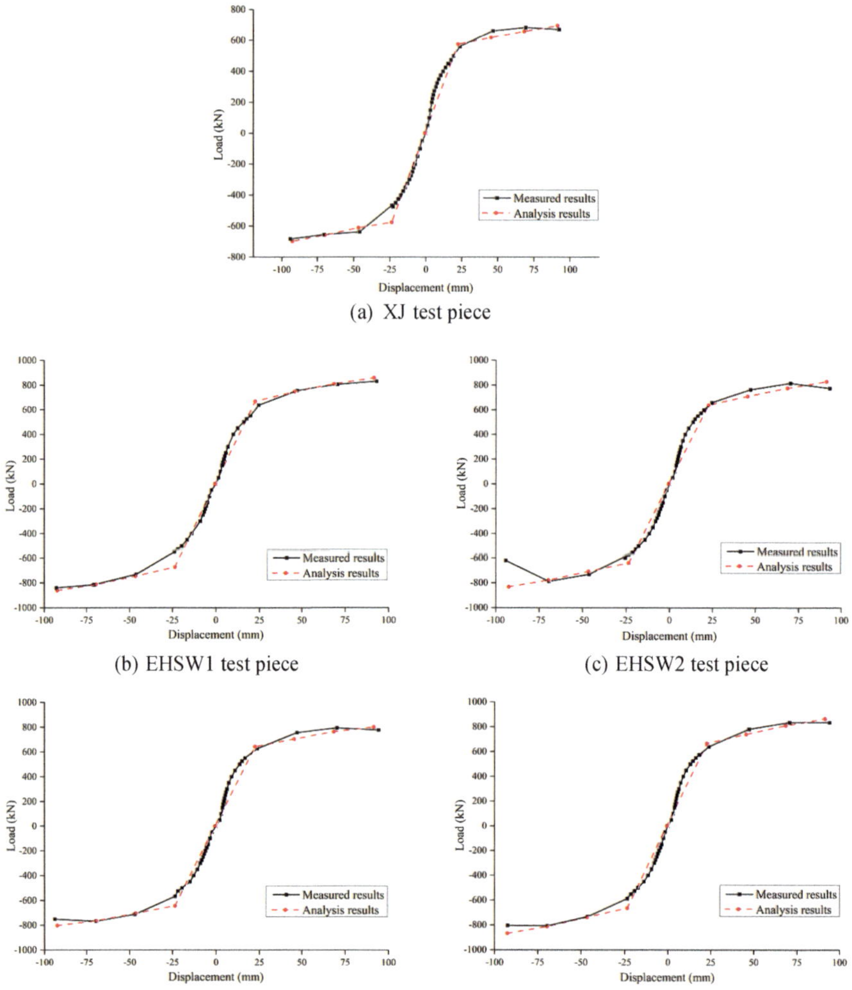

(a) XJ test piece

(b) EHSW1 test piece

(c) EHSW2 test piece

(d) EHSW3 test piece

(e) EHSW4 test piece

Fig. 7.8 Comparison of skeleton curves

the related inherent influence decreases, so that the subsequent numerical simulation curve and the test curve show a good approximation effect. Through the above comparison, it can be found that the difference between the finite element model and the actual component performance is small, which can be used as a basis for subsequent mechanical properties research.

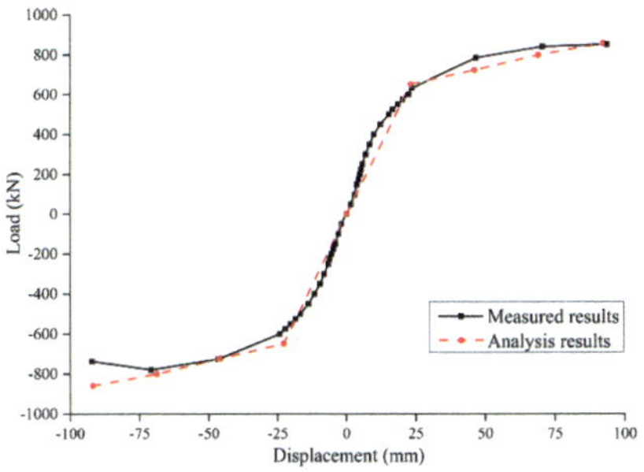

(f) EHSW5 test piece

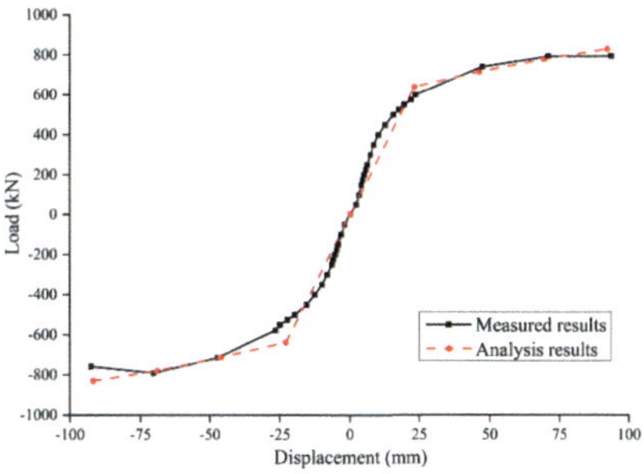

(g) EHSW6 test piece

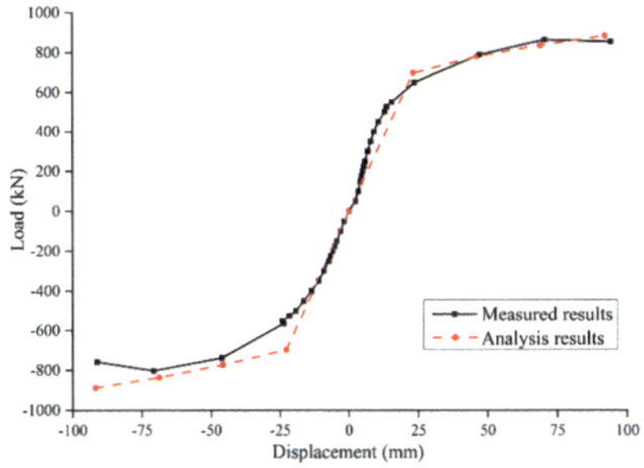

(h) EHSW7 test piece

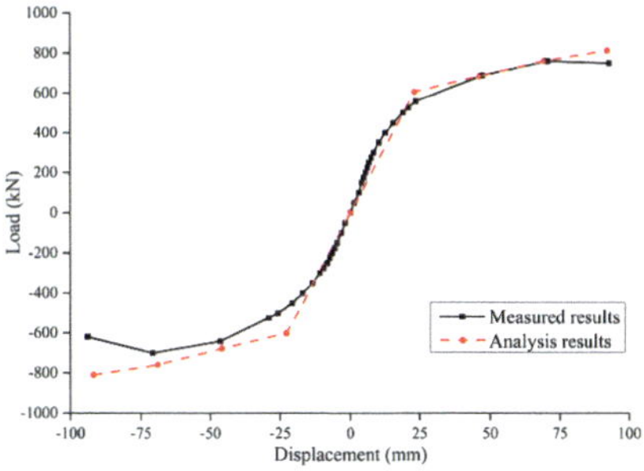

(i) EHSW8 test piece

Fig. 7.8 (continued)

7.3 OpenSees Analysis of Model Construction Improvement

Based on test results and ABAQUS parameter analysis conclusions, the optimal design scheme for EHSW has been obtained: the number of prestressed tendons is 3; the pre-tension is 416.6 kN; the unbonded length of slurry anchor steel is 150 mm, but with the EHSW structure remaining unchanged, still following the 'equivalent cast' concept. To further improve the seismic performance of EHSW, following the concept of fully assembled shear wall, the influence of different structural improvement measures is discussed, such as the position of prestressed reinforcement, the type of prestressed reinforcement, the bonding condition of prestressed reinforcement, and the connection scheme of vertical distributed reinforcement, all relative to the optimal design of EHSW. The more efficient and reliable optimal structure for EHSW is finally determined.

To simplify comparison, the constitutive material parameters for all improved models and optimized EHSW are recalibrated using design values specified in 'Code for design of concrete structures' (GB 50010–2010) [4]. The calculation process for the constitutive model curve is no longer described. A comparison of skeleton curves is provided.

7.3.1 Position of Prestressed Tendon

This paper discusses the influence of the location of unbonded prestressed reinforcement on the seismic performance of EHSW. In the first improvement scheme, the prestressed steel bars are located at the center of the wall section and the center of the concealed column area at both ends, with one bar arranged everywhere. In the second improvement scheme, the prestressed steel bars are positioned in the center of the dark column area at both ends, with two bars set at each end. In the third improvement scheme, the position of the prestressed steel bar is the same as in the second scheme, with the difference being the number of prestressed steel bars, which is changed to one. The pre-tension of a single prestressed steel strand in each scheme is 139 kN. The above three improvement schemes are referred to as EHSW-G1, EHSW-G2, and EHSW-G3, respectively. In addition to changes in prestressed steel bar position and quantity, other conditions are the same as in the previous optimal design EHSW.

The section diagram of each improvement scheme (with the section taken from the bottom slurry anchor area) is shown in Fig. 7.9.

The simulation results of each model are shown in Fig. 7.10 and Table 7.2.

The hysteresis curve and skeleton curve show that the shape of the hysteresis loop for each improved model is basically similar, and overall they are reversed 'S' type.

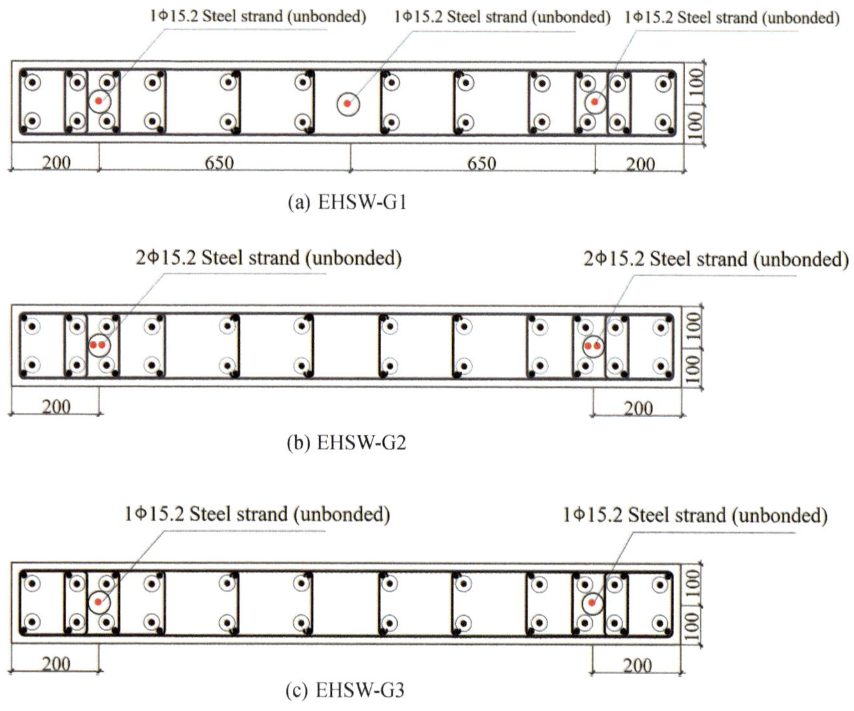

Fig. 7.9 Section diagram of prestressed tendon position improvement scheme

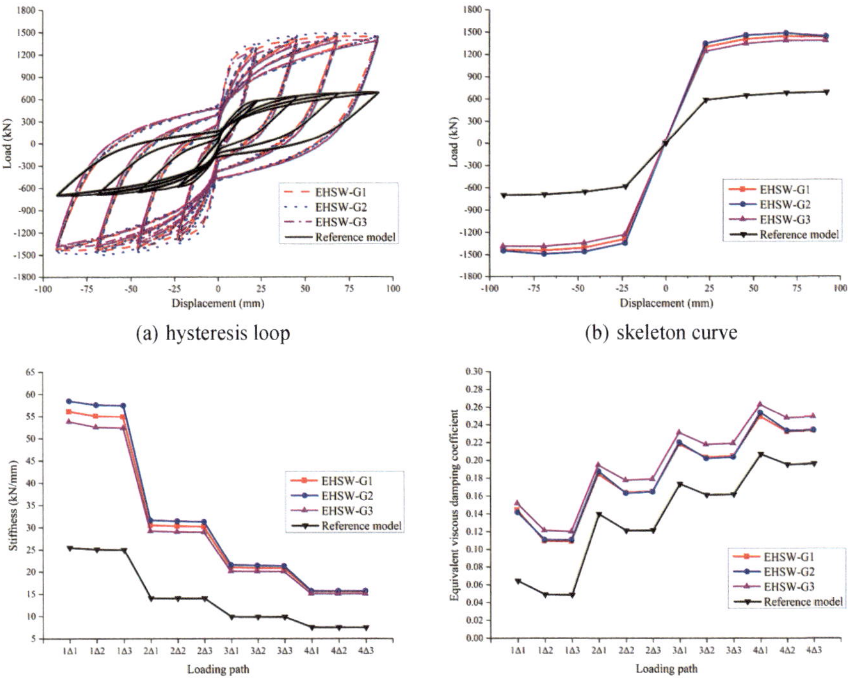

(a) hysteresis loop (b) skeleton curve

(c) stiffness degradation curve (d) equivalent viscous damping coefficient

Fig. 7.10 Calculation result curve of prestressed reinforcement position improvement scheme

Table 7.2 Calculation data of prestressed steel bar position improvement scheme

Model types	Yield load/kN	Peak load/kN	Equivalent yield displacement/mm	Displacement ductility factor
EHSW-G1	1302.77	1441.57	25.71	3.58
EHSW-G2	1351.49	1484.76	25.41	3.62
EHSW-G3	1247.67	1383.40	25.76	3.57
Reference flowsheet	597.09	692.91	27.19	3.38

Note In this chapter, the yield displacement in the calculation formula of displacement ductility coefficient is determined by Park method. Since the load of each model does not decrease significantly during the simulated loading process, the final loading displacement is taken as the ultimate displacement

The trend of the skeleton curve for each model is basically close. After yield, the curve continues to rise slowly, indicating that the bearing capacity of the specimen still has a certain degree of improvement. After approaching the peak load, the curve is basically flat and there is no obvious sudden drop in bearing capacity. This shows that the model can still maintain sufficient bearing capacity under large displacement in the later stage of loading, and has good ductility.

Moving the unbonded pre-stressed reinforcement increases the internal force arm of the pre-stressed reinforcement, thus making full use of the bending resistance of the pre-stressed reinforcement and greatly improving the bearing capacity of the model. The peak load of the EHSW-G1, EHSW-G2, and EHSW-G3 models is increased by 108%, 114%, and 99% respectively compared to the reference model.

Given a constant pre-stress control stress of prestressed reinforcement, the yield load of EHSW-G1 is 118% higher than that of the reference model, the equivalent yield displacement is reduced, and the displacement ductility coefficient is increased by 5%. At the same time, it can be found that keeping the prestressed steel bar arrangement at the center of the concealed column unchanged at both ends, the total pretension increases from 278 kN of EHSW-G3 to 556 kN of EHSW-G2, the yield load significantly increases, the equivalent yield displacement reduces, the corresponding displacement ductility coefficient slightly increases, and the model's deformation capacity is enhanced.

From the stiffness degradation curve, it shows the three improved models' stiffness is significantly higher than the benchmark model, indicating adjusting the prestressed steel bar position (i.e., in the dark column area) can greatly improve the initial stiffness of the model. The stiffness degradation trend of each model is roughly the same, with EHSW-G1, EHSW-G2, and EHSW-G3 gradually reducing the stiffness improvement effect compared to the benchmark model as displacement loading increases, but they remain above the benchmark model's stiffness degradation curve. In the later stage of large displacement loading, the stiffness degradation curves of the three improved models tend to overlap, showing similar stiffness characteristics.

The position of prestressed reinforcement also significantly influences the equivalent viscous damping coefficient. Each improved model's energy dissipation capacity is notably enhanced compared to the benchmark model. Among them, the energy dissipation capacity of the EHSW-G1 and EHSW-G2 models is essentially equivalent. With increasing loading displacement, the EHSW-G3 model performs better, floating above the first two improved models' curves. The analysis indicates that the EHSW-G3 has lower prestressed tendon pretension than EHSW-G1 and EHSW-G2, advancing the steel bar yield. Under the same ultimate displacement, the yield degree of the slurry anchor steel bar is improved, enhancing the yield energy dissipation of the steel bar and improving the model's energy dissipation capacity to some extent.

In general, the three improved models' seismic performance indexes are close, significantly enhanced compared to the benchmark model, demonstrating good seismic performance. Considering actual construction, the second improved scheme with slightly better performance can be adopted to achieve more ideal seismic performance.

7.3.2 Types of Prestressing Tendons

The influence of changing the type of prestressed steel bar on the seismic performance of EHSW is discussed. Based on the optimal design of EHSW, the original prestressed steel bar type is replaced by 1860 grade Φ15.24 prestressed steel strand to 930 grade prestressed rebar with a diameter of 32 mm. The improved scheme is referred to as EHSW-G4, and the other conditions are the same as the previous optimal design EHSW.

The simulation results are presented in Fig. 7.11 and Table 7.3.

From the skeleton curve, stiffness degradation diagram, and equivalent viscous damping coefficient curve, as well as Table 7.3, it can be seen that the change of

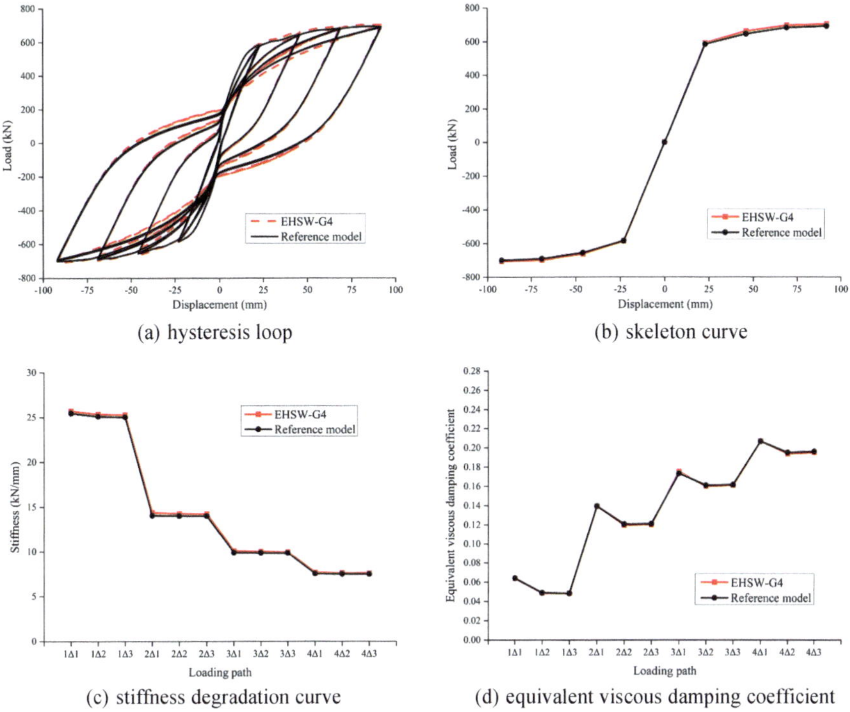

Fig. 7.11 The calculation result curve of prestressed tendon type scheme

Table 7.3 Calculation data of prestressed tendon type scheme

Model types	Yield load/kN	Peak load/kN	Equivalent yield displacement/mm	Displacement ductility factor
EHSW-G4	603.75	702.17	27.35	3.36
Reference flowsheet	597.09	692.91	27.19	3.38

Note In this chapter, the yield displacement in the calculation formula of displacement ductility coefficient is determined by Park method. Since the load of each model does not decrease significantly during the simulated loading process, the final loading displacement is taken as the ultimate displacement prestressed reinforcement type has little effect on the bearing capacity, energy dissipation, and displacement ductility coefficient of the model, indicating that the change of prestressed reinforcement type has little effect on the seismic performance of EHSW, and the two have the same seismic performance.

Therefore, in actual construction, to facilitate construction, prestressed steel strands can be directly considered for application.

7.3.3 Prestressed Tendon Bonding Situation

The influence of prestressed steel bar adhesion on the seismic performance of EHSW is discussed. Based on the optimal design EHSW, the unbonded prestressed steel bar is changed into a bonded prestressed steel bar, and the other conditions are the same as the optimal design EHSW. This scheme is referred to as EHSW-G5.

The simulation results are shown in Fig. 7.12 and Table 7.4.

The hysteresis curve shows that the pinching effect of the model is weakened after changing to full-length bonded prestressed reinforcement, residual deformation increases, and plastic deformation of the structure increases.

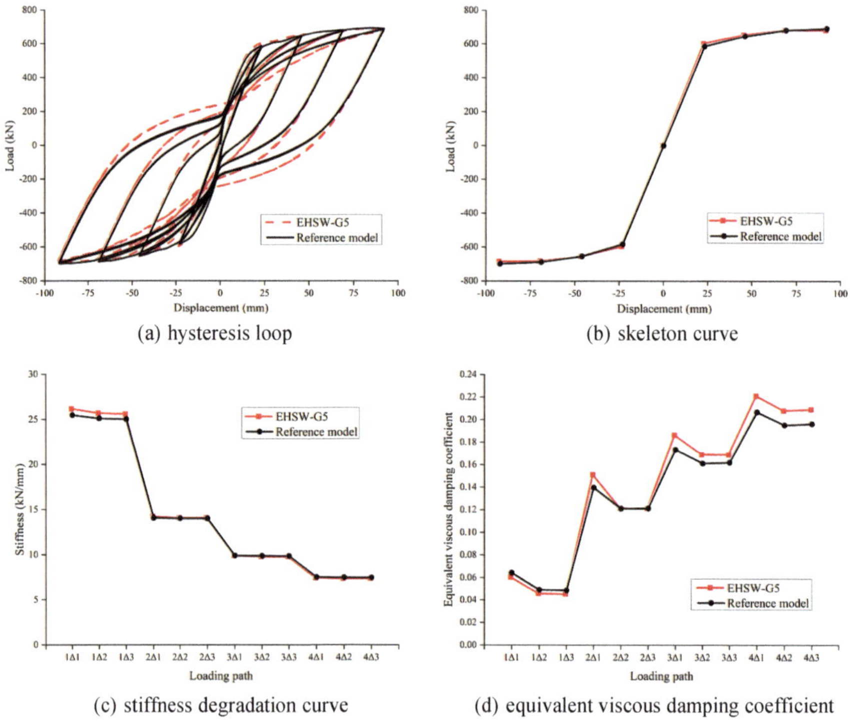

(a) hysteresis loop (b) skeleton curve

(c) stiffness degradation curve (d) equivalent viscous damping coefficient

Fig. 7.12 Calculation result curve of prestressed tendon adhesion

Table 7.4 Prestressed tendon bonding calculation data

Model types	Yield load/kN	Peak load/kN	Equivalent yield displacement/mm	Displacement ductility factor
EHSW-G5	608.98	682.11	27.06	3.37
Reference flowsheet	597.09	692.91	27.19	3.38

It can be seen from the skeleton curve and Table 7.4 that although the peak load of the EHSW-G5 model is slightly lower than that of the benchmark model, the yield load and displacement ductility coefficient are basically the same as the benchmark model, indicating that the deformation capacity is basically the same.

It can be seen from the stiffness degradation curve that the stiffness of EHSW-G5 from the beginning of loading to the 2Δ displacement cycle is slightly higher than that of the reference model, indicating that the overall strength of the concrete increases after the change to the full-length bonded prestressed steel bar, which effectively improves the stiffness of the model at the initial stage of loading. From the 2Δ displacement cycle, the local damage of the bonded prestressed concrete gradually increases, and the stiffness of the EHSW-G5 model gradually decreases in the middle and late stages of loading. The stiffness degradation curve basically coincides with the reference model, reflecting the same stiffness characteristics.

It can be seen from the equivalent viscous damping coefficient curve that the energy dissipation capacity of EHSW-G5 is slightly lower than that of the reference model at the initial stage of loading, but the energy dissipation capacity of EHSW-G5 begins to exceed that of the reference model from the beginning of 2Δ displacement cycle. With the increase of loading displacement, the energy dissipation capacity of EHSW-G5 gradually increases, and it floats above the reference model curve after the beginning of 3Δ displacement cycle. This shows that the damage of bonded prestressed concrete in the middle and late stages increases significantly, which improves the energy dissipation capacity of the model.

In general, the seismic performance of the two is basically the same. In actual construction, the combination of the two can be used to achieve more ideal seismic performance.

7.3.4 Vertical Distribution Steel Bar Connection Scheme

The paper discusses measures to improve the connection of vertical distribution steel bars, with improvement directions divided into two categories. The first is to adjust the diameter and quantity of vertically distributed steel bars in the middle area of the wall to minimize the workload of steel bar connection. The second category is to consider directly canceling the connection of vertically distributed steel bars.

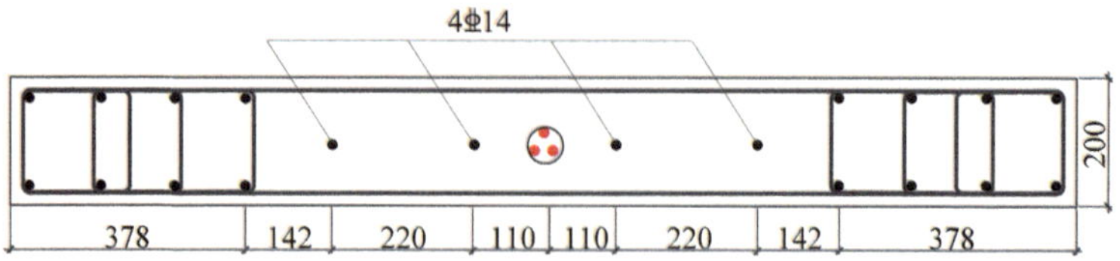

(a) EHSW-G6

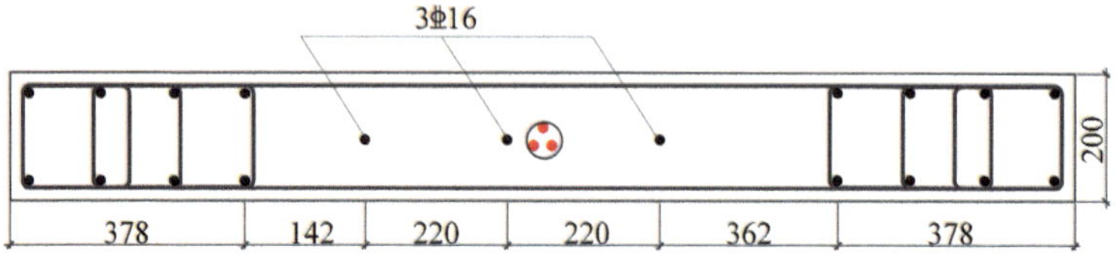

(b) EHSW-G7

Fig. 7.13 Section diagram of improvement scheme for adjusting vertical distribution steel bar

(1) Adjust the diameter and number of vertical distribution steel bars

In this improvement scheme, 4C14 and 3C16 are used to replace 8C10 in the middle area of the optimized design EHSW wall, with the replacement steel bars placed at half the section thickness. All other conditions remain the same as in the original optimized design EHSW, referred to as EHSW-G6 and EHSW-G7.

The specific schematic diagram of the improvement scheme (section taken from the upper non-slurry anchor area of the wall) is shown in Fig. 7.13.

The simulation results are shown in Fig. 7.14 and Table 7.5.

The hysteresis curve, skeleton curve, stiffness degradation curve, and Table 7.5 show that the pinch effect, bearing capacity, stiffness, and displacement ductility coefficient for EHSW-G6, EHSW-G7, and the benchmark model are nearly identical, indicating that the change in vertical distribution reinforcement in the shear wall (excluding concealed column areas) has little effect on the seismic performance of EHSW.

The equivalent viscous damping coefficient curve shows that the curves for EHSW-G6, EHSW-G7, and the benchmark model have the same trend. Specifically, the curves for EHSW-G7 and the benchmark model are almost coincident, with the EHSW-G6 curve floating slightly below, indicating that the energy dissipation capacity for EHSW-G7 and the benchmark model is equivalent, while EHSW-G6 is slightly weaker. Overall, the three models are basically on the same energy consumption level.

Therefore, in actual construction, the EHSW-G7 model with less reinforcement can be chosen, which facilitates construction and saves steel while achieving the same seismic performance as the original reinforcement.

(2) Cancel vertical distribution of steel connection

The central area of the wall (excluding the two ends of the hidden column section) of the grouted anchor steel bar is removed, that is, only the bottom area of the shear wall

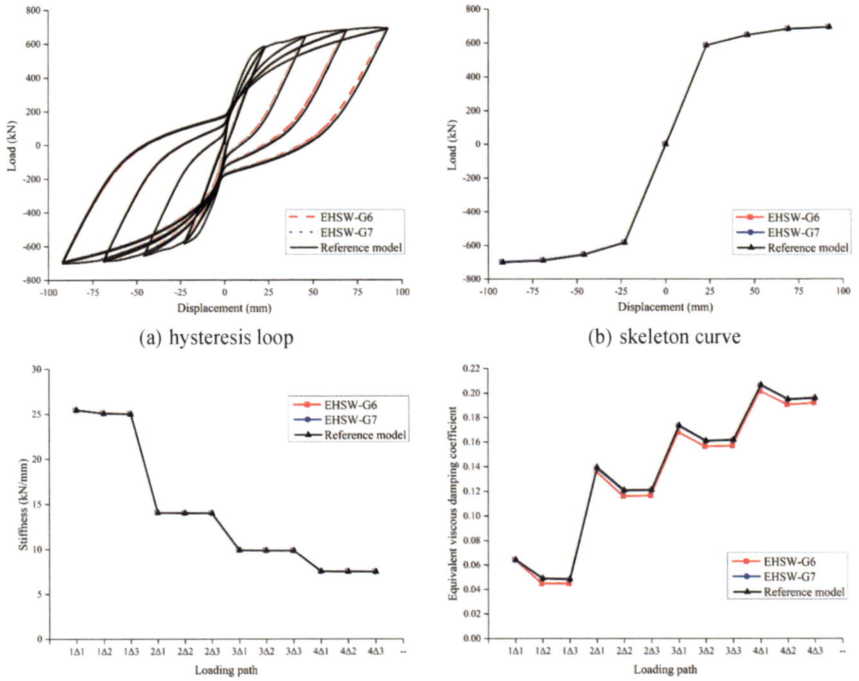

(a) hysteresis loop

(b) skeleton curve

(c) stiffness degradation curve

(d) equivalent viscous damping coefficient

Fig. 7.14 Adjust the vertical distribution of steel bar diameter and quantity scheme calculation results curve

Table 7.5 Adjust the vertical distribution of steel bar diameter and quantity scheme calculation data

Model types	Yield load/kN	Peak load/kN	Equivalent yield displacement/mm	Displacement ductility factor
EHSW-G6	596.61	692.39	27.20	3.38
EHSW-G7	596.16	691.33	27.18	3.38
Reference flowsheet	597.09	692.91	27.19	3.38

has grouted anchor steel bar in the hidden column section, and there is no vertical steel bar bonding between the bottom seam and the base in the central area. The remaining conditions are the same as the original optimization design EHSW, and the improved scheme is referred to as EHSW-G8.

The schematic diagram of the improvement scheme (with a section taken from the bottom grouted anchor area) is shown in Fig. 7.15.

The simulation results are shown in Fig. 7.16 and Table 7.6.

The hysteresis curve shows that the disconnection of ordinary steel bars at the bottom joints of the middle area of the wall has a slight influence on the pinching

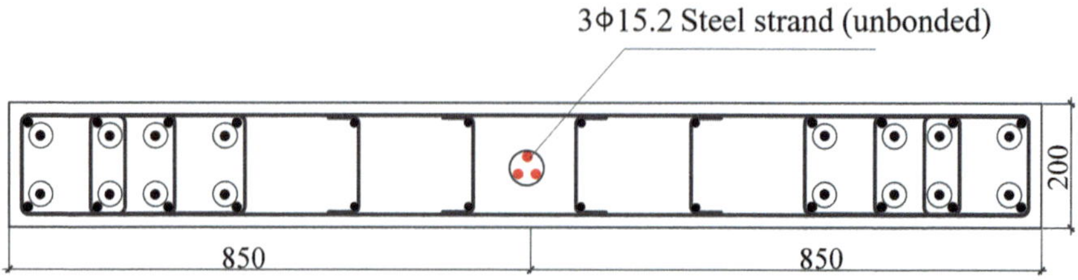

Fig. 7.15 Cross-section diagram of canceling vertical distribution steel bar connection scheme

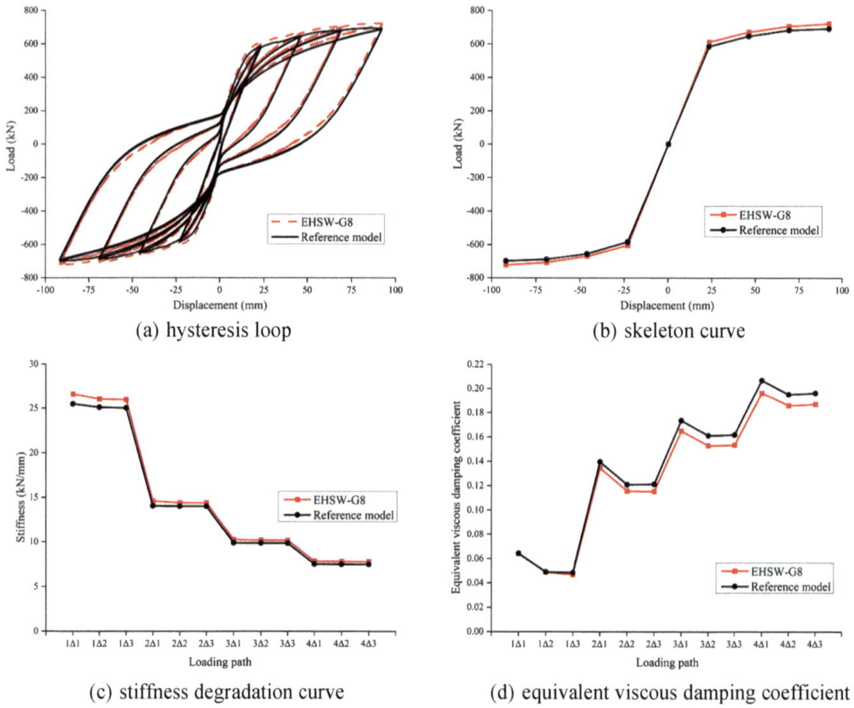

Fig. 7.16 Calculation result curve of canceling vertical distribution steel bar connection scheme

Table 7.6 Calculation data of canceling connection scheme of vertical distributed steel bar

Model types	Yield load/kN	Peak load/kN	Equivalent yield displacement/mm	Displacement ductility factor
EHSW-G8	620.85	722.56	27.06	3.40
Reference flowsheet	597.09	692.91	27.19	3.38

effect of the whole model, slightly enhancing it, and slightly reducing the residual deformation of the model.

From the skeleton curve and Table 7.6, we can see that the peak load of the EHSW-G8 model is 4.5% higher than that of the benchmark model, indicating that the disconnection of the ordinary steel bar at the bottom joint of the middle area of the wall and the base remaining unbonded slightly improves the bearing capacity of the structure.

The yield strength of the EHSW-G8 model is 4% higher than that of the benchmark model, and the equivalent yield displacement and displacement ductility coefficient are basically the same as the benchmark model, indicating that the deformation capacity of the two is basically the same.

The initial stiffness of the EHSW-G8 model is slightly higher than that of the reference model, but with the increase of the loading level, the difference between the two models decreases and gradually remains stable. The stiffness degradation curve of the EHSW-G8 model is close to that of the reference model in the middle and later stages of loading, showing similar stiffness characteristics.

From the equivalent viscous damping coefficient curve, we can see that the energy dissipation capacity of the EHSW-G8 model from initial loading to 2Δ displacement cycle is equivalent to that of the benchmark model. Since there is no ordinary steel bar in the middle of the shear wall to participate in energy dissipation, the energy dissipation capacity of the 2Δ displacement cycle is lower than that of the benchmark model. With the increase of displacement loading level, the gap between the benchmark model and the EHSW-G8 model gradually becomes larger, but the overall numerical difference remains at a low level. This shows that the vertical distribution of steel bars in the middle of the wall is discontinuous, which reduces the energy dissipation capacity of the model to a certain extent, but the overall energy consumption of the model is not significantly weakened, still meeting the seismic performance requirements. □

7.3.5 Determination of Structural Improvement Scheme

Based on the optimal scheme obtained from the improved structural analysis above, an optimal design scheme is proposed to achieve the best seismic performance of the EHSW system. Combined with the above analysis and comprehensive consideration, the following improvement scheme is obtained. The vertical steel bar in the middle area of the wall (except for the dark column area at both ends) is 3C16, the steel bar is placed in half of the thickness of the wall section, the spacing of the steel bar is 220 mm, and the vertical distribution of the steel bar in the middle area of the wall is not connected. The prestressed steel bar adopts 1860 Φ15.24 prestressed steel strand, and two bars are set in the central area of the hidden columns at both ends. The whole is in a long unbonded state, and the total pretension is 556 kN. This improvement scheme is referred to as EHSW-G9.

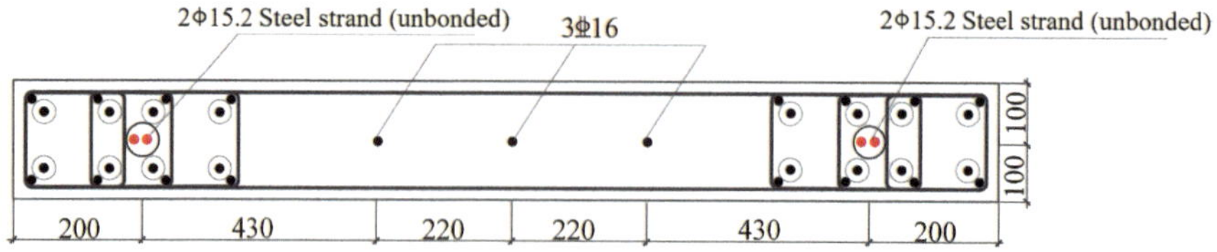

Fig. 7.17 Structure improvement scheme section diagram

According to the above calculation and analysis results, it can be found that the comprehensive seismic performance of EHSW-G2 is the best among the optimal schemes in the previous improved structures. Therefore, EHSW-G2 is selected as the control model of the optimal scheme in this section.

The schematic diagram of the construction improvement scheme (with the section taken from the bottom slurry anchor area) is shown in Fig. 7.17.

The simulation results are shown in Fig. 7.18 and Table 7.7.

The hysteresis curve shows that the shape of the hysteresis loop for each model is basically similar, overall forming a fuller reverse ‘S’ type. The pinching effect

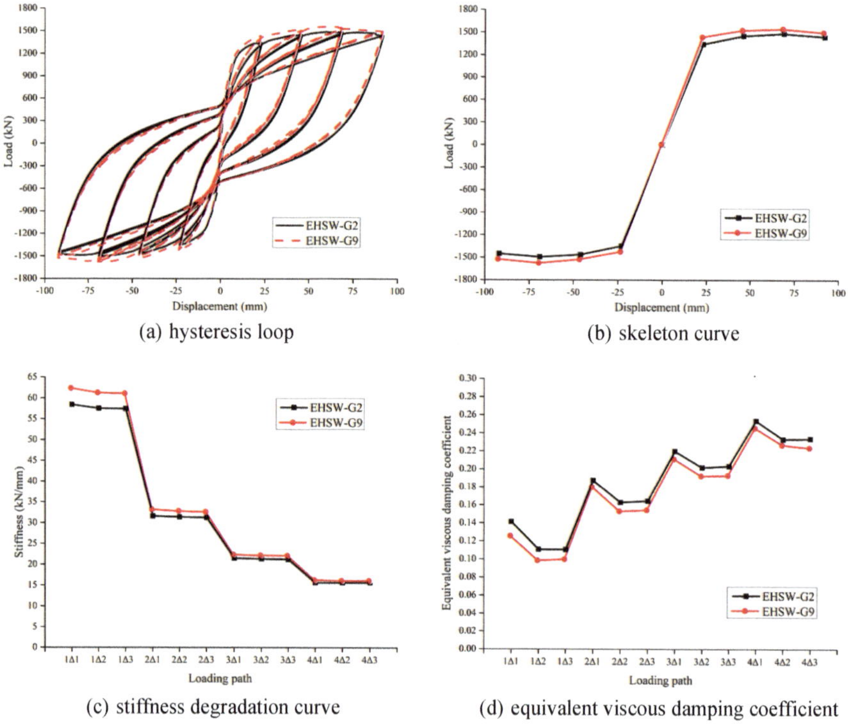

(a) hysteresis loop
(b) skeleton curve
(c) stiffness degradation curve
(d) equivalent viscous damping coefficient

Fig. 7.18 Calculation result curve of improved construction scheme

Table 7.7 Calculation data for construction improvement scheme

Model types	Yield load/kN	Peak load/kN	Equivalent yield displacement/mm	Displacement ductility factor
EHSW-G9	1440.74	1542.73	24.74	3.72
EHSW-G2	1351.49	1484.76	25.41	3.62

for EHSW-G9 is better than for EHSW-G2, indicating that the residual deformation for the model is slightly reduced. Analysis shows that the self-centering ability for the equivalent model has been slightly improved due to disconnection of the reinforcement at the bottom joint in the middle of the wall.

The skeleton curve and Table 7.7 show that the bearing capacity for EHSW-G9 is slightly higher than for EHSW-G2, with peak and yield load increased by 4% and nearly 7%, respectively, and equivalent yield displacement and displacement ductility coefficient improved accordingly, indicating that the deformation capacity for the model is enhanced.

The stiffness degradation curve shows that the initial stiffness for EHSW-G9 is significantly higher than for EHSW-G2. As displacement loading level increases, the stiffness improvement effect gradually decreases, and the two show similar stiffness characteristics.

The equivalent viscous damping coefficient curve shows that the equivalent viscous damping coefficient for each model gradually increases with displacement, indicating that model damage constantly develops. Compared with EHSW-G2, the equivalent viscous damping coefficient for EHSW-G9 decreases, that is, the hysteresis curve fullness decreases. Analysis shows that due to slight reinforcement decrease in the middle area for EHSW-G9 and complete disconnection between the base joint reinforcement and base, the unbonded state is maintained, which is equivalent to direct energy dissipation member reduction. Therefore, in the displacement loading process, plastic deformation is smaller, which also delays reinforcement yield in the hidden column area to a extent, so that the model energy dissipation capacity is slightly reduced.

In summary, the improved model seismic performance is generally better than for EHSW-G2, and it is finally determined as the optimal design structure, that is, the section configuration scheme shown in Fig. 7.17.

References

1. Mckenna F, Fenves GL. Open system for earthquake engineering simulation (OpenSees) user manual. Pacific Earthquake Engineering Research Center, University of California, Berkeley. http://opensees.berkeley.edu, 2006.
2. Taucer FF, Spacone E, Filippou FC. A Fiber beam-column element for seismic response analysis of reinforced concrete structures. Berkeley: UCB/EERC-91/17; 1991.

3. Scott BD, Park R, Priestley MJN. Stress-strain behavior of concrete confined by overlapping hoops at low and high strain rates. ACI J. 1982;79(1):13–27.
4. National standard of the People's Republic of China. Code for design of concrete structures GB 50010–2010 (2015 Edition). Beijing: China Architecture & Building Press; 2015.

GPSR Compliance

The European Union's (EU) General Product Safety Regulation (GPSR) is a set of rules that requires consumer products to be safe and our obligations to ensure this.

If you have any concerns about our products, you can contact us on ProductSafety@springernature.com

In case Publisher is established outside the EU, the EU authorized representative is:

Springer Nature Customer Service Center GmbH
Europaplatz 3
69115 Heidelberg, Germany

Batch number: 10704307

Printed by Printforce, the Netherlands